The Infinite Cosmos

Questions from the frontiers of cosmology

The Infinite Cosmos

Questions from the frontiers of cosmology

Joseph Silk

OXFORD

UNIVERSITY PRESS

OXFORD
UNIVERSITY PRESS

Great Clarendon Street, Oxford OX2 6DP

Oxford University Press is a department of the University of Oxford.
It furthers the University's objective of excellence in research, scholarship,
and education by publishing worldwide in

Oxford New York

Auckland Cape Town Dar es Salaam Hong Kong Karachi
Kuala Lumpur Madrid Melbourne Mexico City Nairobi
New Delhi Shanghai Taipei Toronto

With offices in

Argentina Austria Brazil Chile Czech Republic France Greece
Guatemala Hungary Italy Japan Poland Portugal Singapore
South Korea Switzerland Thailand Turkey Ukraine Vietnam

Oxford is a registered trade mark of Oxford University Press
in the UK and in certain other countries

Published in the United States
by Oxford University Press Inc., New York

© Joseph Silk 2006

British Library Cataloguing in Publication Data
Data available

Library of Congress Cataloging in Publication Data
Data available

Typeset in Minion
by RefineCatch Limited, Bungay, Suffolk
Printed in Great Britain by
Clays Ltd., St. Ives plc

ISBN 0–19–850510–8   978–0–19–850510–5

1

To Jonathan, Timothy and Jonah

◼ CONTENTS

■ LIST OF FIGURES

1   Introduction

Two things are infinite – the universe and human stupidity, and
I'm not sure about the former.

Albert Einstein

The past! the infinite greatness of the past! For what is the
present after all but a growth out of the past?

Walt Whitman

The infinite has to be a relative concept. Go any distance: an infinite space
means that there is more to explore. The end is never reached. Could the
universe be like that?

The notion that the universe could be infinite has excited extreme
views. To some, the concept of an infinite universe is horrifying. The
chances of creating intelligent life may be infinitesimal, but they are finite.
It happened at least once! In an infinite universe replicas of us are inevitable. There would be identical copies of ourselves replicating every action,
every word, somewhere. This seems hard to accept. But this is hardly a
good scientific reason for abandoning an infinite universe. Indeed,
modern astronomy tells us that the infinite universe may be more than
just a metaphysical concept. Physics may even require the universe to be
infinite. And if it were actually infinite, would we ever know? I shall argue
that there are means of exploring both a very large but finite and conceivably even an infinite universe. If the universe is topologically small – that
is, very large but finite in volume – there may be clues to be unlocked in
such observables as the fossil radiation left over from the Big Bang. In
contrast, exploration and verification of an infinite universe is not far
from the voyages of science fiction, but arguably lies in the realm of
physics.

What is certain is that the universe is very, very large. Even if it is not
actually infinite, it is likely to be far, far larger than the remotest horizon
visible with the world's most powerful telescopes. At the very least, the

universe is nearly infinite. In this book, I will describe how astronomers have come to this conclusion and how the concept of an extremely large but finite universe is testable by experiment.

2 Perspectives

If the doors of perception were cleansed, everything would appear to man as it is, infinite.

William Blake

We are all in the gutter, but some of us are looking at the stars.

Oscar Wilde

Humanity is a mote in the eye of the universe. Man has been a distinct species for one-thousandth of the age of the universe. The fragility of life bodes ill for the future. The most minuscule of cosmic impacts between the Earth and a wandering asteroid could readily erase all traces of life on Earth.

But we have survived so far, over billions of years. That is no mean accomplishment, and one that inevitably prompts some hard questions. By what circuitous route did we get here? As a cosmologist, I go as close to the beginning of the universe as theory allows. Whatever my theory, there is always a point of departure. Our knowledge of physics only takes us back so far. Before this instant of cosmic time, all the laws of physics or chemistry are as evanescent as rings of smoke. Speculation runs rife, among theologians and philosophers as much as among astrophysicists.

In the beginning there was the realm of theology. But there also is the realm of philosophy and of cosmology. All attack the beginning of the universe from very different perspectives. For example, a philosopher might say, 'OK, if there is a beginning, there was something before the beginning. Otherwise the notion of a beginning makes no sense.' But the scientist (or the theologian) says, 'By definition, what happened before is not science (or theology).' The new cosmology starts out from a beginning, at an instant when there was nothing. Or as close to nothing as makes any difference. Nothing that is in the form of matter. There was space, but it was empty. Time did not exist. Hence the instant that we refer to as the beginning may have lasted a very long time indeed, long enough for strange things, apparent violations of mass and energy conservation, to occur. We are all taught at school that energy is conserved. But the quantum theory allows us to borrow energy, as long as we pay it back before we can be observed. Our understanding of the beginning of the universe relies on the theory of the fundamental nature of matter as well as on observations by astronomers of the most distant reaches of the universe.

The art of cosmology

> What I am going to tell you about is what we teach our physics students in the third or fourth year of graduate school. . . . It is my task to convince you not to turn away because you don't understand it. You see my physics students don't understand it. . . . That is because I don't understand it. Nobody does.
>
> Richard Feynman

Cosmology is the study of the universe: its structure, its beginning, its fate. The cornerstone of modern cosmology is the Big Bang theory. But how much of this is mere speculation? And what are the hard facts about the beginning of the universe? I will recount what we really know about the Big Bang theory. This will involve such issues as evolution on a grand scale: from the genesis of matter to the birth and death of entire galaxies of stars. Defining the difficult questions is the first step towards finding the answers. Armed with knowledge, the ultimate questions about the universe are accessible. Even if we cannot answer them, there is much to be learned in the attempt.

Galaxies are far away. Light from them has travelled for millions of years, and as we study them we look into the past. Probing the past has changed our views dramatically. In the middle of the seventeenth century Archbishop James Ussher of Ireland made the startling revelation that God created Heaven and Earth on October 22, 4004 BC, at 8 o'clock in the evening! This was later revised by the English biblical scholar Dr John Lightfoot, who gave the date for the creation of Adam as October 23, 4004 BC, at 9 o'clock in the morning.

A geologist digs deeper and deeper to probe the geological past. The oldest rocks on the earth are about three billion years old. He or she will even find older non-terrestrial rocks, meteorites, the oldest rocks in the solar system, that are 4.6 billion years old.

But the cosmologist can do better, simply by looking out into space. The centre of our Milky Way is about 20,000 light years away. The Andromeda Galaxy, our nearest neighbour of comparable size and a naked eye object, is two million light years away. We see it as it was at the dawn of man. The most distant galaxies known are more than ten billion light years away. Their light was emitted long before the sun and solar systems had formed, and even before our galaxy was created. The universe is a big place. But it is not so big that we cannot try to understand its origin, its evolution. Creation can be viewed with the aid of modern telescopes. The Big Bang is our modern, scientific understanding of creation. It replaces Zeus and Thor, Adam and Eve. But unlike myth, the scientific story is based on evidence. Not quite the evidence you can hold in your hand, but certainly the evidence you can see through a telescope.

The building blocks of cosmology

> The difficulty lies, not in the new ideas, but in escaping the old ones, which ramify, for those brought up as most of us have been, into every corner of our minds.
>
> John Maynard Keynes

Galaxies are the building blocks of the Universe, hence I begin with them. We are surrounded by such star clouds, once called 'island universes': indeed, we inhabit one, the Milky Way galaxy. A hundred billion suns orbit its uncharted space, mostly confined to a thin disk that extends for tens of thousands of light years but has a thickness of merely a few hundred light years. The stars are fossils of a glorious past, when galaxies shone a hundredfold more brilliantly than they do today. Astronomers have acquired immense amounts of data that chart the stars that populate the innards of galaxies over almost the entire electromagnetic spectrum, from radio waves to gamma rays. Relatively few secrets remain, and we understand, at least to a first approximation, the intimate secrets of the galaxies. We can track the evolution of the Milky Way back in time, to an epoch long before the Sun had formed.

I will use galaxies as probes of time itself, as a means of reaching back towards the Big Bang. We can see giant galaxies far away, and their birth lit up the dark spaces of the remote universe. To paraphrase Isaac Newton in his comment on a rival, Thomas Hooke, a man of reputedly dwarfish appearance, 'we are like dwarfs who can see immensely further by standing on the shoulders of giants'.

Galaxies take us a long way back into the past. In later chapters, I will describe the universe before galaxies had formed. This is where speculation runs rife, but it takes us inevitably towards the very beginning, and beyond, the realm of the ultimate questions.

3   **Principles**

You can't learn too soon that the most useful thing about a
principle is that it can always be sacrificed to expediency.

Somerset Maugham

The only principle that does not inhibit progress is: anything
goes.

Paul Feyerabend

The cosmologies of myth were characterized by the invention of deities to
account for phenomena, such as the stars or the planets. There was one
law for the heavens, where the gods dwelt for the most part, and another
for the ordinary man. The modern scientist, and we can look back to
Aristotle as being the first great natural scientist, commences his study of
the universe by assuming that the laws of nature are universal. We do not
need one law for the planetary motions, and another for walking along a
road. In this spirit, cosmology, the scientific study of the universe, is
developed by extrapolating locally verified laws of physics to remote locations in space and time. We do not need Atlas to hold up the heavens, but
the orbits of the planets suffice to stop them falling on to the Earth like so
many shooting stars. At first, crystalline spheres undergoing complex
arrays of epicyclic motions were needed to account for the planetary
motions. It took nearly 2000 years for the Copernican revolution to over-
throw the epicycles of the geocentric universe, dethroning the Earth as the
centre of the cosmos and setting the stage for modern astronomy.

The Big Bang theory provides a framework for studying our past and
for predicting our future. The Big Bang is the modern scientific version of
the creation story. It is a theory and it is a model that can be firmly
anchored in the context of modern observations. Above all, it is a simplification of the astronomical complexities of nature that reveals an elegant
underlying symmetry and beauty. In the best traditions of science, it is a
highly predictive theory, and one that has been systematically refined as
the observational database grows.

Simplicity

Consider first the simplification aspects of the theory. Look at a deep image of the sky, taken with a large telescope. The universe is teeming with galaxies. There is no doubt that our local patch of the universe is highly lumpy. Yet take a step backwards. Compare images taken in different directions. As long as one steers well clear of the Milky Way, the universe looks much the same wherever we look. This leads us to a simplifying conclusion. Imagine smoothing out the images, by, for example, peering at them through an out-of-focus lens: the details are degraded. Imagine inspecting the images through a filter of frosted glass: all structure is lost. But the general pattern of light, due to the galaxies, remains. The universe now seems nearly uniform and the same in all directions. In effect, one is filtering the observed structure, and only retaining information on structure that is larger than that allowed by the filter. Approximate uniformity appears once the universe is smoothed over a few million light years.

Once one smears over structures such as galaxies and galaxy clusters, the distant universe is approximately uniform or homogeneous. The *cosmological principle* asserts that the universe is approximately the same in all directions. We cannot say look over there and find the universe is much hotter than over here. We say the universe is statistically isotropic. There is no direction one can point towards and say: the centre of the universe is over there, where the density of galaxies seems to increase. Of course, these observations are made from our vantage point. The cosmological principle also requires that the universe should be approximately uniform. No steep gradient in density should be seen, as might be expected if we were at the centre of a very large hole in the distribution of galaxies. The cosmological principle generalizes the appearance of homogeneity and isotropy to observers situated anywhere in the universe. The universe is said to be homogeneous for any observer.

One motivation behind the cosmological principle is the need to dethrone us as being privileged observers from the vantage point of the Earth. The universe is assumed to be, on the average, isotropic at all times for sets of fundamental observers. One consequence is that the universe must, on the average, also be homogeneous.

If ever it turns out that local physics is inadequate for describing the universe, then we are at liberty to develop new physical laws that reduce to generalizations of local physics. The great successes of modern physics, such as the generalization of Newton's gravity to general relativity and the standard model of particle physics, are characterized by simplicity. Such considerations about the simplicity of a successful theory can be incorporated into a simple cosmological principle that serves as a guide for building a model of the universe. Ultimately, we have to verify the cosmological principle experimentally.

Principles are the bread and butter of cosmology. Perhaps it all began with Plato, who had rather fixed ideas about the underlying rules that the universe should satisfy. He wanted the planets to follow perfect circles, and his ideas held sway for nearly 2000 years until Kepler, by applying the data painstakingly accumulated by Brahe, proved that they were wrong. The lesson is of course that empirical data must come first, but the further moral is that they can only be understood with the aid of fundamental laws or principles. Newton, and then Einstein, demonstrated this for planetary orbits.

The starting point for the Big Bang theory is Einstein's theory of general relativity in combination with the cosmological principle. General relativity is a theory of gravity that starts with the concept that space is measurable, whatever the scale. This need not be so, and indeed certainly is not so on infinitesimally small scales. But we do not need to worry about the exotic phenomena of the quantum theory. General relativity has now been verified to high precision via observations of some remarkable astronomical objects, the binary pulsars. These are pairs of orbiting neutron stars. These compact objects are so close that their orbits decay by emission of gravitational waves, a prediction of the theory of general relativity. Two of these systems are being monitored.

The cosmological principle has also been verified to exquisite precision, in so far as this can be achieved from our vantage point. Here the major breakthrough has come with the exquisite smoothness of the cosmic microwave background, and hence of the density of the early universe.

The echo of the Big Bang, the faint glow we call the 'cosmic microwave background', has demonstrated homogeneity and isotropy to a level of the order of 1 part in 100,000. Homogeneity has also been demonstrated by galaxy surveys that provide three-dimensional maps of the universe. From the spectrum of the light of a galaxy, we infer that the galaxy is never at rest, but generally is receding from us, with the light being shifted to longer, or redder, wavelengths than for nearby galaxies. From the redshifts, one infers the distances. And they are enormous, as inferred given the strong empirical correlation discovered by Edwin Hubble in 1929 between redshift and distance. We will discuss this relation below in more detail.

As one probes deeper and deeper into the universe, to distances as great as several billions of parsecs (1 parsec = 3.2 light years), the density of galaxies is found to be uniform. We definitely do not inhabit a universe of vanishingly low density in the mean, as some have argued. One can set a limit on any large-scale non-uniformities of around 10 per cent.

We learn from Copernicus

The cosmological principle reduces to the statement that the universe, on the average, looks the same from any location. There is no preferred direction. We cannot prove the cosmological principle: it is a matter of philosophy or even faith, albeit faith that is highly motivated.

The cosmological principle is motivated by the Copernican argument that the Earth is not in a preferred position at the centre of the universe. Why should the Earth be special? This is an unlikely beginning point for a theory that is based on objective facts about nature. More to the point, any geocentric theory is horribly complicated. Ptolemy constructed a geocentric universe with a multitude of crystal concentric spheres, rotating on different axes, and centred on the earth. By the time of Copernicus, the number of transparent spheres invoked to explain the planetary motions had multiplied beyond belief, in the theory of epicycles. The resulting mathematics was ugly. Copernicus sliced most of the crystal spheres away to construct a far simpler and more elegant heliocentric cosmology, centred on the Sun. Astronomers, centuries later, confirmed that the Earth moves in orbit around the Sun.

But why stop at the Sun? Why not place the entire universe on a similar footing? There need be no preferred location for the Sun, nor for any galaxy such as our own. Imagine viewing the universe from any point in space. If from each point there is no preferred direction, the universe is locally isotropic, and it must also be uniform in space. The cosmological principle states that the universe is approximately isotropic and homogeneous, as viewed by any observer at rest.

This version of the cosmological principle is the foundation stone of modern cosmology. How can we verify such a concept? The astronomers of the early decades of the twentieth century had the universe firmly planted in and around the Milky Way galaxy. The notion of an expanding universe would have been completely heretical. The next major advance in cosmology was to require a paradigm shift driven by data. But in order to provide the infrastructure to do any modern cosmology at all, the cosmological principle provided the essential foundation. The evidence was at first rather sketchy. Half a century had to elapse before modern observations provided robust confirming evidence. Observations of the cosmic microwave background have vindicated the cosmological principle. Einstein originally applied the cosmological principle in 1916 to infer that the universe was static. After all, at that time, there was little believable evidence in favour of a systematic expansion of space. Indeed, this seemed completely counter-intuitive to people educated with a Cartesian view of the world.

Within half a century, the cosmological principle became a fact. A stronger version, the perfect cosmological principle, goes further: the universe appears the same from all points and at all times. In other words, there can have been no evolution: the universe must always have been in the same state, at least as viewed over a sufficiently long time. In this sense, the perfect cosmological principle contrasts with the weaker version, which allows the possibility of very different past and future states of the universe. The perfect cosmological principle spawned a theory in which the appearance of the universe never changed, despite its expansion. Matter came from literally nowhere to replace the expanding matter and maintain a constant average density. This theory became known as the steady state universe, which postulated that matter was continuously being created, a concept and a cosmology that has long been discredited. We see the dramatic evolution of the universe as we look far away, and consequently back in time.

Universality

> This result is too beautiful to be false; it is more important to have beauty in one's equations than to have them fit experiment.
>
> Paul Dirac

Physicists have long believed that the physical laws that govern the evolution of the universe should be simple. In the eyes of a physicist, simplicity is beauty. This is not necessarily true for an artist, of course. Simplicity in physics is an overriding requirement. If the theory has too many bells and whistles, it dies a natural death. This was the case with the Platonically inspired theory of planetary motions, based on epicycles. Circles, considered to be a perfect form, were the inspiration for a complex hierarchy of epicycles that allowed the planets to orbit a stationary Earth. Even before the observational evidence was compelling, astronomers, most notably Copernicus, became increasingly dissatisfied with the geocentric Ptolemaic model of the solar system.

The starting point of modern cosmology, even before adopting the cosmological principle, is to postulate that the laws of physics are universal in space and time. If physical laws vary in random ways from one location to another, we might as well return to mythology for an equally compelling model of the universe. While one can test the simplest derivations of the laws of physics, such as possible variations in the gravitational constant and the fine-structure constant over a very limited range of space and time, remote variations in extreme regions of space and time are untestable. Of course, at the singularity predicted at the origin of the Big Bang, our laws of physics may break down. However, this surely is symptomatic of an inadequacy in our theory. One hope for further progress is that some day we will discover the Ultimate Theory of Everything. Perhaps this will be able to cope with the beginning of the universe in a self-consistent way. Such a theory may be none other than the theory of quantum gravity, the long-sought union between Einstein's theory of gravitation and the Planck-inspired quantum theory. With unlimited freedom to imagine variations in the fundamental constants and the laws of physics, any cosmology is in danger of entering the realm of metaphysics, something unverifiable and beyond science. Hence we usually prefer

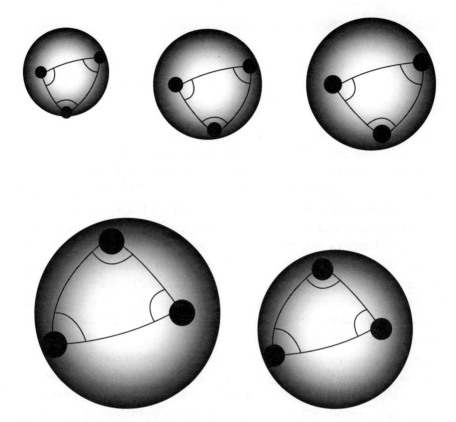

Figure 1 A general triangle stays the same shape as the universe expands only if the universe expands at the same rate in all directions and if it expands at a velocity that increases linearly with the length of any side of the triangle. If, however, the expansion velocity were to increase – say, as the square of the length of a side – the triangle would soon change shape.

to assume that, apart from considerations of the initial singularity, which lasted a mere 10^{-43} seconds, the laws and constants of physics are unchanged.

Unlike other branches of science, cosmology is unique in that there is only one universe available for study. We cannot tweak one law of physics, juggle another, and end up with a different system on which to experiment. We can never know how unique is our universe and the physical laws that describe it, for we have no other universe with which to compare. The universe denotes everything that is or ever will be observable by us.

Are we essential to the universe?

> We are bits of stellar matter that got cold by accident, bits of a
> star gone wrong.
>
> Arthur Eddington

Cosmologists have fertile imaginations. We can imagine other possible universes. One could have a universe in which conditions were so turbulent that no galaxies formed. There would be no stars and no planets. Needless to say, man could not exist in such a universe. The very fact that our species has evolved on the planet Earth sets significant constraints on the possible ways our universe has evolved.

Such an anthropomorphic approach may be the only way we can ever tackle such questions as: why does the proton have a mass that is much larger (precisely 1836 times larger) than the electron, or why is the neutron just 0.14 per cent heavier than the proton? If these mass ratios were very different, we certainly would not be here. The idea that our actual existence sets certain constraints on physical laws and on the nature of the universe has been enshrined into a principle: the anthropic cosmological principle. This states that our universe must be congenial to the origin and development of life. This is why we are observing our particular universe. Others may or must exist, but are lifeless and unobserved.

The vastness of the universe makes one think that there ought to be life out there somewhere. Cosmologists can explain the enormity of the universe, and indeed predict that it is immensely vaster than anything we can see, by postulating that there was once a brief moment of greatly accelerated expansion. This phase of inflation, as it is called, has dramatic consequences. The great vistas of space open up.

At the onset of inflation, the universe was only a tiny fraction of a second old. Ordinarily, light would only travel a fraction of a light second since the Big Bang. But inflation allows space to expand exponentially and thereby removes the barrier for light propagation. Light now has time to travel millions or billions of light years, even though the universe was a tiny fraction of a light second across at the onset of inflation. There is no horizon to the propagation of light that limits our cosmic vista. The propagation barrier has fallen. This is not magic, nor does it involve faster-than-light travel. It simply involves a short cut through space that becomes accessible during the moment of inflation.

Imagine a universe in which the brief phase of inflation, which many cosmologists swear by and which accounts for the vastness of the universe, has somehow gone awry. Normally, inflation also accounts for the structure of the universe. It grasps fluctuations on the tiniest of scales, those of the quantum, and stretches them to enormous distances. Microscopic quantum fluctuations in density are boosted up to macroscopic scales and act as seeds around which galaxies form. But perhaps the inflation-generated seeds were far too sparse and weak. Galaxies would not have formed in such a universe, nor would stars, nor planets. There would be no astronomers to observe the universe. And what, conversely, if the fluctuations from the beginning were far too strong? Black holes would have formed in enormous numbers. They would outnumber the stars. There could still be planets – after all, planets have been found in such unlikely places as in orbit around neutron stars. So it might be difficult to exclude life forms. But the universe would be a very different place from the one we observe.

What bizarre universes! Ones that we clearly have successfully evaded. But why? One answer is that the universe is somehow tailored to our requirements: the just-so universe. The Big Bang was big, but not too big nor too small; it was hot, but neither too hot nor too cold; it was inhomogeneous, but not too clumpy nor too smooth. This is the solution offered by the Goldilocks principle, or, under its more scientific name, the anthropic principle. Of all the possible universes that do exist, life can exist in only an infinitesimally small subset. The fact that we are here selects this subset of possible universes; the rest may or may not exist, but go their own way, unobserved.

Of course, there is no physics behind such reasoning. It appeals to the luck of the draw in a cosmic lottery. The anthropic cosmological principle explains nothing, and is devoid of any fundamental significance. It cannot discriminate between a flea and an elephant, let alone man. The anthropic cosmological principle argues that the universe must have been constructed so as to lead to the development of intelligence, in order for observers to be present. This version of the cosmological principle begs the question of how likely is the development of life, and indeed the evolution of the observable universe, from suitably random initial conditions. This should be an issue resolvable by physics rather than by fiat.

Notwithstanding the need for physics input, many cosmologists have been tempted by the ease with which anthropic arguments can provide answers to such seemingly imponderable questions as: why is this universe the way it is, why are we here now, why are we here at all, or why these laws of nature? According to the British cosmologist Brandon Carter, there are many otherwise unexplained coincidences in nature that conspire to allow life to be possible. These are all necessary conditions for life. But Carter went a step further. For him, the only synthesis was that, sooner or later, intelligent life was inevitable.

Scientists, especially cosmologists, love principles. After all, the cosmological principle carried Einstein far, if at first in the wrong direction. Einstein did not realize that his equations allowed expansion, and his application of the cosmological principle led him to postulate that the universe is static and supported by the antigravity associated with the cosmological constant he introduced (we will hear more about the cosmological constant later).

The anthropic principle asserts that the universe is just so because we are here. If it were any different, there would be no cosmologists to observe it. One can now supplement the cosmic garden with this cosmic principle, and our very own Big Bang emerges. One can understand, so we are told, why the universe is so vast and relatively uniform, and why it is just beginning to undergo a phase of acceleration away from the Big Bang.

We seem to be in good shape. Never mind that the most powerful minds in physics, so they tell us, have been working for two decades on a theory of elementary particles that posits the existence of elongated higher-dimensional structures called superstrings within which the particles and their interactions are incorporated. Superstring theory is mathematically elegant, and even compelling, although we have yet to see a single unambiguous and experimentally verifiable prediction. Certainly, it is a phenomenally difficult theory. Is the anthropic principle part of the solution? I do not wish overly to dampen the party spirit, but I am underwhelmed.

From physics to metaphysics

> I believe there are 15,747,724,136,275,002,577,605,653,961,
> 181,555,468,044,717,914,527,116,709,366,231,425,076,185,
> 631,031,296 protons in the universe and the same number of
> electrons.
>
> Arthur Eddington

Great physicists have a tendency to delve into metaphysics in their later
years. Indeed, some go far astray from physics. Sir Arthur Eddington was
an eminent astronomer who spent the last two decades of his life wrestling
with, and even publishing a monograph on, an ultimately incomprehensible fundamental theory of physics. He thought he could predict the
value of the fine structure constant.

The anthropic principle is essentially metaphysics, and that is the essence
of the problem for most physicists in accepting it. The anthropic logic is
either immensely subtle, by arguing that we, via our mere existence, control the cosmos, or unabashedly naive, by setting aside any physics explanations that any ultimate theory of physics might reasonably be expected to
deliver. Metaphysics lacks predictive power, the very core of physics. The
anthropic principle is an unabashed expression of our ignorance.

Here is a simple argument that demolishes the anthropic principle.
Suppose one were convinced that in order for life to emerge, armed with
consciousness and intelligence, one required the fundamental constants of
physics to have certain values. There is one especially important constant
that controls all of chemistry, the fine structure constant. This is measured
to be 1/137, and not, say, 1/150. One can show, were the fine structure
constant 1/150, that carbon molecules would never have assembled,
carbon nuclei would never have formed, and a vital ingredient of life
would be missing. Is this a vindication of anthropic reasoning? On the
contrary, it points to a fatal flaw, because experiments find that the fine
structure constant is 1/137.03599911. Such accuracy is far beyond any
anthropic reasoning or logic. The explanation of the value of this constant
must lie elsewhere.

It may well be that the ultimate theory of cosmology will have
anthropic ramifications. We are some way yet from this promised land.
However, most cosmologists prefer to believe that there are underlying

physical theories, of which we are still largely ignorant, that tune the universe to provide a more welcoming environment. There could be a host of radically different universes that we need not worry about. An initially chaotic universe, with different conditions and even different fundamental particles, may have evolved via the physics of quantum gravity into setting the scene for the beginning of our universe.

The power of gravity

Gravity is the weakest of the fundamental forces of nature. It worked for Isaac Newton as, fleeing Cambridge from the plague in 1665, he passed his days speculating about why the Moon orbited the Earth, finding his legendary revelation reputedly when the apple fell. Compared to inter-atomic forces, gravity is weaker by 40 powers of 10. Electric charges cancel out, as atoms are neutral in charge. The electromagnetic force exerted by the Sun on the Earth is negligible.

Gravity, though, is another matter. The atoms act in concert to contribute to the Sun's gravitational attraction for the Earth, and for the Earth on a falling apple. Gravity plays a role on the Earth only because there are some 10^{80} atoms in the Earth, all pulling in the same direction. Newton showed that gravity flattened the rotating Earth, much to the consternation of the French astronomers. Led by Jean-Dominique Cassini and his son Jacques, the viewpoint from Paris was that the Earth was shaped like an egg, as experiments allegedly demonstrated. Newton won this debate, although it took a century for the dust to settle.

Newton went on to invent calculus, although the Germans assigned priority to Gottfried Wilhelm Leibniz. The last word in gravity did return to Germany as Albert Einstein, based in Berlin from 1914 to 1933, supplanted Newton's theory of gravity with the theory of general relativity that he announced in 1916. According to Einstein's geometric theory, gravity operates by curving space, and particles of matter follow the trajectories of curved space. Sir Arthur Eddington eagerly jumped upon the new theory. He mounted an expedition, as soon as the First World War ended, to observe the total solar eclipse of 1919. The results produced banner headlines across the world.

The general theory of relativity predicts that the deflection of light from a star viewed near the sun's disc – only measurable during an eclipse – would be twice that predicted by the original corpuscular theory of light in flat space. Eddington's observations showed that Einstein was right: space is curved.

Not only space but also time is distorted by gravity. Clocks slow down. The frequency of light emitted by a source in a gravitational field is lowered, as viewed by a distant observer. The effect that one infers on the surface of the Earth is small, only amounting to about one part in a billion. However, the effect has been measured in a series of experiments pioneered at Harvard University. A physics laboratory was constructed with no nails in the floor or walls in the interest of avoiding weak magnetic effects that could have perturbed the highly accurate measuring devices that were used. A difference in the frequency – a tiny gravitational redshift – was found over a height of 22 metres using gamma ray lines from a radioactive isotope of iron that could provide frequency measurements to an accuracy of 1 part in 10^{15}.

Time dilation turns out to have practical implications. One is in the programming of cruise missiles. Timing with the global positioning satellite system (GPS) is used to compute the trajectories of missiles. Without the inclusion of the orbital correction from general relativity, a cruise missile on a transcontinental flight would miss its target by a kilometre or more. The attraction of gravity is inescapable, although rarely fatal. It guides our understanding of the cosmos.

4 Our Neighbourhood

The men of experiment are like the ant, they only collect and use; the reasoners resemble spiders, who make cobwebs out of their own substance. But the bee takes the middle course: it gathers its material from the flowers of the garden and field, but transforms and digests it by a power of its own. Not unlike this is the true business of philosophy (science); for it neither relies solely or chiefly on the powers of the mind, nor does it take the matter which it gathers from natural history and mechanical experiments and lay up in the memory whole, as it finds it, but lays it up in the understanding altered and digested. Therefore, from a closer and purer league between these two faculties, the experimental and the rational (such as has never been made), much may be hoped.

Francis Bacon

Look up at the sky on a dark night. The thousands of twinkling points of light that you can see are mostly stars that are neighbours of the sun. Neighbourliness is a relative concept. Here it amounts to a distance of hundreds of light years. Stars come in many varieties, from red to blue, and even beyond the range of colours visible to the human eye. The bluer the colour, the hotter the star.

The distance to the nearest stars is determined by parallax. During the course of a year, as the Earth orbits the Sun, the apparent position of a star on the celestial sphere is displaced appreciably. The displacement depends on its distance from us. The parallax method can be used to measure stellar distances out to a hundred light years or so. Knowing the distance and the apparent brightness gives the luminosities of the stars – how bright they actually are, or how much energy they are emitting in a given time. Our Sun is pumping out radiation (and energy) at a rate of some four hundred million trillion megawatts. By comparison, the largest power stations on the earth have power outputs that amount to tens of megawatts. Sadly, most of the solar power is wastefully radiated away into empty space.

Astronomers also measure the colours of the stars. Some stars are red, some are blue, and some are yellow, like our Sun. In effect, the astronomer is measuring the distribution of light with wavelength, or the spectrum of the star, in order to obtain a precise colour. Red stars are brighter at long wavelengths, blue stars are brighter at short wavelengths, relative to each other. From the colour or the spectrum, the temperature of the surface of the star can be inferred. This is the effective temperature, the light energy that leaves the surface of the star. For example, the Sun has an effective temperature of 6000 kelvin. (Physicists measure temperatures on the absolute temperature scale, in kelvin (K), with 0 K ('absolute zero', the lowest possible temperature) corresponding to -273 degrees Celcius. One kelvin is the same as $1°C$.) The coldest stars are about 2000 K, and the hottest stars are about 50,000 K. Their colours span the spectral range from infrared to ultraviolet; the Sun is a yellow star. The interior of a star will of course be much hotter, under the enormous pressure of the great ball of gas that constitutes a star.

Now the luminosity and the temperature are related by the surface area of the star. At a given temperature, the larger the surface area, the more luminous is the star. We infer that most stars are about as large as the Sun. Some are much larger, veritable giants, and much more luminous than the Sun. Some are much smaller, mere dwarfs, and much dimmer than the Sun. Our nearby stellar population is dominated by its run-of-the-mill citizens, with the occasional giant and dwarf interspersed. The typical stars are burning hydrogen as their nuclear fuel. Giants have exhausted their hydrogen cores, and are beginning to burn helium. Dwarfs have mostly exhausted their thermonuclear fuel supply, and are shining on borrowed time.

Stellar evolution theory tells us that what distinguishes normal hydrogen burning stars from one another is their fuel supply. More fuel means more mass, and more mass means more gravity. In particular, more mass generates more gravitational force at the surface of the star, and the pressure and temperature at the centre of the star are higher. The result is that massive stars are hot and luminous, and are relatively short-lived. They burn more and more brightly as helium is burnt, and end up, briefly, as giants, as their outer layers expand. Most of the mass is shed in this high luminosity phase in a wind, and what is left of the star finally shrinks down to a dwarf when no more fuel remains to be burnt. Stars with a mass similar to that of the Sun are several billions of years old, while the most massive stars burn brightly and are fated to die within millions of years.

Star death is heralded by a transient giant or even supergiant phase. Consider the star Betelgeuse. This red giant has swelled up to the size of our solar system, out to beyond the orbit of Jupiter. Expansion causes cooling, so the largest giants are red. On the other hand, contraction causes heating, so dwarfs start off as blue or even ultraviolet, only to fade into blackness after billions of years.

The bigger picture

Most of the nearby stars around us are moving with the sun, orbiting the Milky Way at a speed of some 200 kilometres per second. The Milky Way is actually a giant disc of stars, about 100,000 light years across. A galactic year is long. The Sun orbits the galaxy every 200 million years. The Milky Way has undergone about 60 revolutions. By analogy to the Earth orbiting the Sun, we can think of the Milky Way as being 60 galactic years old. It has been in existence for about 12 billion years. The oldest stars, which are in the globular star clusters that are found in a spherical region around the disc of the Milky Way, are nearly 12 billion years old, and must have contained some of the first stars to form. The Sun is a middle-aged star, having been born about 4.6 billion years or 23 galactic years ago, in a middle-aged galaxy. The sun is about half the age of most of the stars in the Milky Way.

Some stars are very young. Scattered around the Milky Way are nebulae where stars are being born today. These gas clouds have condensed out of the interstellar medium and eventually acquire enough mass by accreting smaller clouds to collapse under their own weight; they are gravitationally unstable. The collapsing clouds break up into dense clumps of gas, which in turn coalesce into stars. The newly formed stars, some of which are massive and luminous, light up the surrounding nebulosity. This placental gas is viewed in reflected starlight. We can peer into the very cores of star-forming clouds by looking at infrared and radio frequencies. Even the darkest clouds can hold no secrets. Nevertheless, precisely why and how a cloud breaks up to form stars of certain masses, and how rapidly it does so, are still a matter of conjecture. The interstellar medium is complex. It is turbulent. It is permeated by magnetic fields, which play a role in its evolution. Predicting what happens to a cloud is more difficult than forecasting the weather, something that we cannot yet do reliably over long periods of time.

Stars age and die, some terminating their days in spectacular explosions. The expanding gas shells, at speeds of thousands of kilometres per second, interact with the pre-existing interstellar gas and debris from earlier phases of stellar evolution. The result is a complex nebulosity, a web of gas shells, sheets, and filaments that permeates interstellar space.

Stellar nurseries are crowded with stars and provide a spectacular view of evolution at work. One of the closest stellar birth sites is in the heart of the constellation of Orion, known as the Trapezium, where thousands of stars are packed into a region that is a few light years across, the distance to our nearest stellar neighbour Proxima Centauri. This is a star cluster. The stars were born together a few million years ago, and the entire gamut of stellar masses is represented. There are stars fifty times the mass of the sun that shine a hundred thousand times more brightly. This profligate rate of energy generation can only be sustained for a few million years, after which these massive stars will explode catastrophically.

There are also stars only a tenth of the mass of the sun, stars that have barely succeeded in igniting their thermonuclear fuel reserves and that shine at a thousandth of the luminosity of the sun. These stars will live for hundreds of billions of years at such a parsimonious rate of radiation. All of these stars shared a common ancestry. About ten million years ago, an interstellar cloud in Orion collapsed and spontaneously spawned thousands of stars as it fragmented. We do not understand the details but the evidence is unambiguously written on the sky.

There are many such regions of star formation scattered throughout the Milky Way. The sites of forming stars lie in a spiral pattern that is created as the Milky Way disk rotates. The pattern is a propagating wave of gas compression. This is rather like the effect seen on the terraces of an old-style football stadium with standing room only, when, as the back of the crowd pushes forward to cheer a goal, the rows in front follow moments later, and the movement is propagated to the front. The galactic gas compression wave is dragged out into a spiral because of the rotation of the disk. Stars form at the crest of the wave. As the wave continues, neighbouring regions are compressed and more stars form. Star formation has been likened to a contagious epidemic as it spreads throughout the galaxy.

What triggers the wave? Probably in many cases it is the close passage of a satellite galaxy that causes a tidal effect, thinning gas as material is tugged under the gravitational attraction of the satellite. The Large Magellanic Cloud fulfils this role for the Milky Way. In other cases, the compression wave is generated by the tumbling of an elongated central core of stars that has a bar-like shape. As the bar tumbles about its centre of mass, it pulls and pushes on nearby stars in the galaxy. As the bar turns, it exerts a tidal force on the galactic disc in which it is embedded. This force varies in strength according to whether the bar is pointing towards or is transverse to a neighbouring region of the disc. The result is that a bar also generates a density wave. The density wave generates the spiral structure. About half of all spiral galaxies have prominent bar-shaped central concentrations of stars, and these are the sources of many of the ubiquitous spiral arms.

The raw material that forms the stars is the interstellar gas. This collects into clouds that aggregate mass as they orbit the galaxy. Eventually the clouds acquire so much mass that they become unstable under their own gravity and collapse to fragment into stars. The star formation process is inefficient. Only 1 per cent or so of the gas forms stars in each rotation period. The clouds reform, and the cycle of starbirth resumes. A galaxy like the Milky Way can continue to form stars for hundreds of rotation cycles before the gas supply is exhausted. There is even a reservoir of additional gas that slowly accretes on to the disk. These are relic gas clouds left orbiting in the halo from ten billion years ago when the galaxy was first contracting. Such infall helps to extend the longevity of the Milky Way.

Eventually the bright lights will dim. The gas that fuels star formation will be exhausted. No more stars will form. The old stars will gradually die. Only the lowest mass, dimmest stars remain. Even these must eventually die. Slowly the Milky Way will fade into oblivion. This will take hundreds of billions of years, but it is inevitable.

The heart of the matter

There is more to the Milky Way than a disc of stars. Even in the 1920s, it was apparent that spherical clusters of stars, the globular star clusters, were satellites of our Milky Way, distributed throughout a gigantic spherical region centered on an obscured patch of sky in the constellation of Sagittarius. It gradually became clear that the centre of our galaxy was in Sagittarius and some 25,000 light years distant on the modern distance scale. The centre was not easily visible because of the obscuring dust in the plane of the Milky Way. The globular clusters mapped out a giant halo 300,000 light years across, and this was thought to demarcate our island universe. The centre of our galaxy was located within the dense spheroidal bulge of stars. Only as infrared observations became routine did the overall pattern of the structure of our galaxy emerge. There was a disc where stars were born and a spheroid of older stars, distinguished by their red colours. The spheroid formed when our galaxy emerged as a distinct entity out of the Big Bang, about 12 billion years ago.

But a surprise lay in store. Five years of repeated observations at infrared wavelengths that could penetrate the galactic dust revealed that stars within a few light years of the galactic centre were moving. Their positions changed slightly every year. After five years their orbits could be traced. That gives us the side-to-side component of the movement. But astronomers could use the Doppler shifts in the spectra of the stars to measure movement towards and away from us too.

The Doppler effect is commonly experienced for sound waves. In a Grand Prix car race, for example, the engine whine of an approaching car reaches a higher pitch, and a receding car a lower pitch, relative to a car that is moving transversely to the observer. The sound waves are Doppler-shifted: to higher frequency or shorter wavelength for approach, to lower frequency or longer wavelength for recession of the emitting source of noise. The same principle applies to light, except that of course the velocities involved are much larger because the speed of light is so much larger than the speed of sound.

Measurement of the Doppler shifts of the stellar spectra enabled astronomers to infer the full three-dimensional orbits of these innermost stars of our galaxy. They were found to be orbiting around the centre of the galaxy at great speed – a speed that only made sense if an extraordinary mass were present at the centre. The required mass surpassed by far all expectations based on the known star density and gravity field associated with the observed stars.

Only one conclusion was possible. There had to be an immense mass of unseen matter lurking within a region a fraction of a light year across. The mass concentration must be so large that even light cannot escape. This is what is known as a black hole. Such objects are predicted to exist by Einstein's theory of general relativity, and stellar mass black holes are known to exist in our own galaxy. But the object in the centre of the Milky Way is exceptional. Only a supermassive black hole could be the culprit, weighing in at some 2.6 million solar masses. We shall see that our Milky Way is far from alone in harbouring such a monster.

The universe of particles

Stars and black holes are formed from atoms, as are people. Our bodies are assemblages of complex molecules that constitute our proteins, DNA, and cells. The molecules consist of atoms of various elements. Carbon, nitrogen, and oxygen are the most important, although traces of iron and other heavy elements are indispensable to the functioning and well-being of our organism. Each atom contains a positively charged nucleus constituted of neutrons and protons, surrounded by a cloud of negatively charged electrons. Most of the atom is empty space: an atom is typically one hundred-millionth of a centimetre across, but the nucleus has a size that is a ten-thousandth of this.

Electrons are fundamental particles. But the heavier neutrons and protons can be broken into smaller parts. When beams of protons collide together at very high energy, the protons are smashed apart. We find that protons and neutrons consist of subparticles called quarks. The quarks have charges of two-thirds and one-third of the charge on a proton or electron. All these particles are collectively called baryons and leptons.

Ordinary matter is made of baryons and leptons. Every human being, every planet, every star consists of baryons and leptons. Most of the mass is in the baryons. A surprise awaits us, however, when we measure the mass of a galaxy, and especially a cluster of galaxies. Baryons no longer suffice. More precisely, luminous baryons, those that we see in stars and in shreds of interstellar and intergalactic gas, do not suffice to account for all of the observed mass. It would seem that most mass is dark. Nor is this the only surprise. While some of the dark mass may be baryonic, most is not. Only about a fifth of the matter in the universe is baryonic. And of these baryons, about 50 per cent are also dark. Approximately twice as many baryons are measured in the early universe as are seen in galaxies. We simply cannot detect the 50 per cent that are dark directly, although their presence is confirmed by the absorption in the spectra of distant galaxies.

All of this has dramatically altered the view of the traditional astronomer, used to the notion that the Milky Way is full of stars, and little more. This has been superseded by modern insights from the interface of astrophysics and particle physics.

A whole new discipline has emerged, called particle astrophysics. The particle physicist rarely has qualms about existence. To her, it is perfectly natural that most of the universe is teeming with exotic particles, interacting so weakly that they are yet to be discovered. To a particle physicist any particle that can exist must once have existed. It may take the most extreme conditions imaginable to create such particles. But such conditions were present in the cauldron of the Big Bang. The particles of which we are made interact with each other, both electromagnetically and by the strong and weak nuclear forces. Electromagnetism controls the world of atoms and molecules, and is responsible for chemistry. The strong interactions hold protons and neutrons together in atomic nuclei, and are ultimately responsible for our stability. But the weak interactions are also important, if more hidden from everyday view. Particles like the neutrino are weakly interacting, and can pass through the Earth unimpeded. Neutrinos are ghostly particles, thought to be massless, and produced in nuclear reactions.

Physicists believe that early in the universe all particles were on an equal footing, regardless of whether today they interact by the electromagnetic, strong, or weak forces. At high enough energy, these fundamental forces all merge together. In fact, one theory, supersymmetry (or SUSY for short), posits that for every known particle there exists a superheavy, weakly interacting, partner particle. Of this vast zoo of particles, few have

5   The Universe of Galaxies

Neutrinos, they are very small.
They have no charge and have no mass,
And do not interact at all.
The Earth is just a silly ball
to them, through which they simply pass
like dustmaids down a dusty hall.
Or photons through a sheet of glass.
They snub the most exquisite gas,
Ignore the most substantial wall.

John Updike

O dark, dark, dark. They all go into the dark.
The vacant interstellar spaces,
the vacant into the vacant.

T. S. Eliot

Galaxies such as our Milky Way system have somehow succeeded in main-
taining a youthful appearance for billions of years. The Sun takes about
100 million years to orbit the Milky Way, and we can think of a solar orbit
as lasting a galactic year. Our galaxy is about 100 galactic years old. By any
measure of the rate at which the gas clouds in the galaxy have collapsed,
there should be little gas remaining. By all accounts it should be a decrepit
assemblage of old and dying stars. Yet it is blossoming with the vitality of
youth, as new stars are born. Indeed, our Milky Way is a fertile breeding
site for stars.

How galaxies remain youthful

The secret to youth is the supply of raw material for forming stars. The gas
is replenished by infall from the surrounding cloud out of which the

galaxy was born. Only a modest amount of infall is needed. Galaxies remain youthful by miserly consumption of their gas supply. Self-regulation explains why galaxies remain gas-rich. Our Milky Way galaxy is in a precarious balance between life and death, between the glories of star formation and the inevitability of stellar senescence.

An elegant theory accounts for the longevity of galaxies. A typical galaxy consists of a spheroidal component of red stars embedded within a disc of blue stars. Red stars are generally cooler stars of low mass and low luminosity, and are long-lived. Blue stars are hot and luminous, and massive. They consume their nuclear fuel rapidly, and are short-lived, lasting some tens or hundreds of millions of years in their bright phase.

It is the discs of galaxies that appear young. Why? The light from discs is predominantly that from the blue, short-lived stars. These stars must form continuously over the lifetime of the disc, which is many billions of years. A disc containing large amounts of gas is unstable. The gas in the disc breaks up under the relentless tug of gravity into clouds of gas. These in turn fragment into stars that populate the disc. Discs of stars are continuously regenerated as gas accretes from the surrounding environment, at least until the reservoir of fresh gas is exhausted. But once the gas is gone, only an ever-dimmer future awaits a galaxy.

Elliptical galaxies contain only old, red stars. These stars formed long ago. Such stars are low mass, similar to the sun, and have a lifetime of ten billion years or more. The stars we see today in an elliptical galaxy formed in the early universe. We conjecture that the event that triggered the formation of the elliptical was the merger of two gas-rich disc galaxies. The net result of two disc-like systems merging is to form a round system, destined to form an elliptical galaxy.

It is tempting to look for patterns in the sky. This after all is how the constellations were named. Scientists classify objects under study into some semblance of order. And classification invariably provides the first clues about origins and evolution.

Galaxies come in three principal varieties. Spheroidal galaxies are red, since they only contain old stars, and are gas-poor. Disc galaxies are blue, because they are forming hot, massive stars, and contain the gas reservoir from which the new stars form. They have prominent spiral arm-like features that demarcate the regions where stars are actively forming. And then there are the misfits, or irregular galaxies.

Edwin Hubble first classified galaxies according to a scheme that contains more loosely wound spirals dominated by giant bulges (Sa galaxies),
as well as tightly bound spirals with minuscule bulges (Sc galaxies). There
were also spheroidal galaxies (ellipticals), disc galaxies without discernible
spiral structure or star formation (S0 galaxies), and of course irregular
galaxies. These have a more complex history that we are beginning to
demarcate with the aid of deep images taken with the Hubble Space
Telescope.

Ground-based telescopes are limited in their ability to resolve fine
structure in galaxy images by atmospheric turbulence, one manifestation
of which is the twinkling of stellar images on a dark night. The typical
image size in good conditions from a large ground-based telescope is
1 second of arc, although images of about half this size are obtained on
exceptionally calm nights. An arc-second is the apparent diameter of a
penny at a distance of one kilometre. Above the Earth's atmosphere, one
can do far better.

With the unexcelled resolution (about one-tenth of an arc-second)
and sensitivity (to twenty-ninth magnitude in the visible, roughly the
amount of light that would be detectable on the Earth from a candle on
the Moon) of the Hubble Space Telescope, one can classify distant galaxies by morphology. As we look at fainter galaxies, the mix of galaxy types
changes. For the brighter galaxies seen in the nearby universe, spirals
dominate. About 30 per cent of galaxies are elliptical galaxies, and 15 per
cent are irregular galaxies. In the distant universe, most galaxies are
irregular. This suggests that galaxy shape or morphology has evolved
with time.

But why has it evolved? The answer is not simple, since irregular galaxies are much less massive than their regular counterparts, but it too has
come from the images taken by the Hubble Space Telescope. By examining
the detailed galaxy images, astronomers have found that close encounters
and even mergers between galaxies become increasingly common as one
looks further back in time. When the universe was half its present age, and
galaxies were closer together, there was at least a 20 per cent probability of
any galaxy undergoing a violent merger with a galaxy of comparable mass.
Today this occurs rarely, about 1 per cent of the time. We infer that
galaxies accumulate mass as a consequence of merging.

A few per cent of nearby galaxies cannot easily be classified. They are

neither fish nor fowl, neither discs nor ellipticals at first sight. Astronomers lump these unclassifiable systems into the rubric of irregular galaxies. Interestingly, the number of irregulars increases markedly as one looks
deeper into the universe. Indeed, the irregulars may be the dominant
population when the universe was a third of its present size.

Why should a typical member of the galaxy distribution in the early
universe be a low mass irregular galaxy? We believe that low mass galaxies
are likely to be irregular because they are more vulnerable to partial
disruption by sustained bouts of star formation and supernova activity.
Massive galaxies by contrast have sufficiently strong self-gravity to develop
a much more regular morphology.

Gas-rich irregular galaxies are typically disc galaxies, but the disc is
masked by the turbulent gas. The turbulence has more than one cause.
Exploding stars stir up the interstellar gas. The gas cools. Now a cold disc
is gravitationally unstable, as pressure forces cannot resist gravity. The
cold gas disc spontaneously fragments into clouds. And not infrequently
there is a collision between a pair of galaxies. Even if one galaxy is much
smaller, the close interaction, which often results in a merger, has a strong
influence on the gas. As for the stars, they are so compact that they rarely
collide, except in the most crowded regions. The gas, however, responds
much more strongly to the changing gravity field, because the gas is
capable of losing energy by cooling.

A merger between two such galaxies is a violent event. The systems are
very responsive to the strong tidal forces operative during the merger. Shock
waves passing through the gas compress and heat it, and gas is disrupted
from the parent galaxies into enormous tidal tails. At the same time, a dense
bar-like structure forms from the stellar cores of the merging galaxies. The
bar exerts strong tidal twisting forces on the gas. The consequence of this
is that the gas loses angular momentum and contracts, and a dense central
mass of gas develops. This soon becomes so massive that it cannot be
stably supported by its self-gravity, and collapse ensues. The gas inevitably
fragments to form stars. The episode of star formation is so intense relative to the prevailing rate of disc star formation that the name 'starburst'
has been given to this phenomenon. Galaxy mergers induce starbursts.

A starburst occurs not over the entire disc of the galaxy, but in the
central few hundreds of parsecs where a mass of gas equivalent to that
of the entire Milky Way interstellar medium has accumulated as a

consequence of the merger. In typical spirals, such as the Milky Way, star formation proceeds at a rate of a few solar masses per year. During a starburst, the star formation rate often exceeds ten or a hundred times that of the Milky Way. More than 90 per cent of this luminosity is absorbed by interstellar dust and reradiated in the far infrared.

The Infrared Space Observatory satellite, launched by the European Space Agency in November 1995 and finally turned off in April 1998, has produced beautiful infrared images of nearby galaxies. This was followed by NASA's Spitzer Space Telescope launched in 2003. The infrared spectral range gives a new and complementary view of galaxies in action. The infrared telescope on board the Infrared Space Observatory discovered many galaxies that were undergoing intense, and necessarily short-lived, bursts of star formation.

There are nearby examples of merger-induced starbursts. Their luminosity is emitted primarily in the far infrared because of the intense concentration of interstellar dust in the central core along with the gas. Indeed, the most luminous galaxies in the universe are inevitably associated with ongoing mergers. The signature of a merger is unmistakable: the tidal tails of ejected stars and gas are clearly visible in deep images.

Worlds in collision

A collision between two galaxies is more catastrophic than the typical high speed car collision. A better analogy might be the momentary close passage of two clouds: a merger is inevitable. The close proximity of the galaxy pair results in strong gravitational forces being exerted on each galaxy. The stellar components respond by merging to form a dense spheroid of stars. The discs of galaxies are destroyed, as stellar orbits become relatively chaotic. The ordered circular motions in the plane of the disc are lost. The gas clouds in the merging discs collide and heat up. Their orbital energy is lost, and radiated away by the gas. Much of the gas falls into the centre of the forming spheroid. Here it collects into a dense cloud that collapses under its own weight. Gas is more responsive than stars to the merger because it is able to lose energy by radiating it away as the gas is shocked.

Did the central spheroids of galaxies form in this manner? The outcome of a merger inevitably is a great concentration of gas in the nucleus. The gas cloud collapses and fragments into huge numbers of stars. Some of these are massive and short-lived. Indeed, so many massive stars are formed that explode as supernovae that there is a back reaction. The exploding stars send out shock waves that accelerate the surrounding gas. Feedback from the accelerated debris with the ambient gas halts the gas collapse. The star formation terminates. Its duration is that of a short, intense burst of activity, lasting tens of millions of years. This is a relatively short duration by galactic standards. The vigorous period of star formation that follows a galactic merger heralds the formation of galactic bulges and elliptical galaxies.

How do we know that this is more than myth? Astronomers have caught merging galaxies in the act of consummation. The close encounter produces dramatic tidal plumes of ejected stars and gas. And some of the gas falls into the nucleus of the galaxy where displays of cosmic fireworks are under way, fuelled by the infall of fresh gas. Such dramatic encounters are rare in the nearby galaxy population. Perhaps one galaxy in a hundred shows evidence of a recent close collision or merger.

But long ago and far away, galaxies were much closer together. Encounters were a frequent phenomenon. Observations show that remote galaxies are smaller than their nearby counterparts, and are often far less regular in shape. All of this attests to the prevalence of mergers in the past. Galaxies began as small gas clouds that aggregated together, forming stars and concentrating more gas, until the grand patterns that we see in mature nearby galaxies eventually developed.

Galaxies change their form, chameleon-like. Typically a galaxy begins as a disc. Gas naturally settles into a disc because of its residual angular momentum. But after a major merger, all bets are off. The disc is likely to be destroyed by the strong dynamical heating and disruptive tendencies of the rapidly varying gravity fields. Gas drains into the centre. The concentration of gas is so strong that a great burst of star formation is inevitable. The result is a dense, spheroidal-shaped distribution of stars, an elliptical galaxy. What happens next depends on the environment. If the galaxy is isolated, gas from the halo rains down gently over billions of years. It settles into a disc. The gas forms stars, and these remain in the disc, on circular orbits in the plane of symmetry. The result is a disc galaxy with a central bulge. Many disc galaxies have bulges, some prominent, others less so. The morphology of galaxies contains a fossil record of their past.

Why was it so very different some ten billion years ago, when galaxies were young? In the age of galactic youth, fuel supplies to supermassive black holes were abundant. The galaxy had far more gas, much of which subsequently formed stars. The reservoir of gas is the ordinary interstellar medium. This was enhanced by a catastrophic event, believed to be the sequel to a collision or a merger with another galaxy, which funnelled gas towards the centre. A merger event is likely to be the consequence of a close encounter between a pair of galaxies. Galaxies that collide are admittedly like ships passing in the night: the stars do not collide. The spaces between the stars are immense enough to guarantee this. However, the close encounter does result in strong gravitational torques being exerted by each galaxy on the other. This results in a loss of orbital energy. The galaxies are destined to merge unless their initial encounter velocity was so large that the ensuing interactions are feeble. Collisions that result in mergers are rare today but were increasingly common in the early universe. Encounters were common in the past, simply because galaxies were much closer together.

The domain of dark matter

Many galaxies have far fewer stars than our Milky Way galaxy. These are dim dwarf galaxies. Most galaxies in the universe are so dim and diffuse as to have eluded detection for many years. Painstaking efforts at chasing elusive hidden evidence for virtually invisible matter were eventually crowned with success. Hydrogen in atomic form is the dominant constituent of the interstellar medium, and it produces radiation at a telltale wavelength of 21 centimetres, in the radio part of the spectrum, where the sky is very dark.

The dark matter breakthrough came with the detection of interstellar gas, gleaming at a radio wavelength of 21 centimetres, and extending to well beyond the peripheries of many galaxies. The dimly glowing gas revealed the presence of vast amounts of hidden mass. The Doppler shift of the radio line emission enabled the mass in stars to be measured. Such

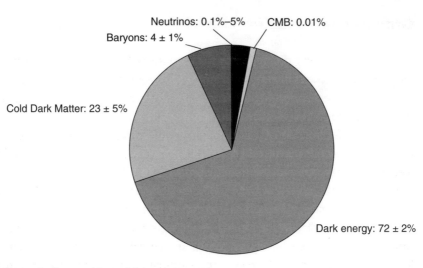

Figure 2 Composition of the universe today.

feebly gleaming collections of stars may be very frequent. Indeed, they may even outnumber the luminous galaxies, such as our nearest neighbour Andromeda, that astronomers have long considered to be the norm. We do not know how many of these low surface brightness galaxies exist, as dim shadows of the grand design spirals that feature prominently in popular catalogues of the universe.

But more than dim stars were discovered. Even with the luminous galaxies, we are just viewing the equivalents of the tops of giant icebergs. Astronomers were surprised and challenged by the realization that 90 per cent of the mass of a galaxy is effectively invisible at any wavelength. Dark matter dominates the outer regions, or haloes, of galaxies. What is the nature of the elusive dark matter?

Particles of ordinary matter are collectively called baryons. These include the protons, neutrons, and electrons from which all atoms and molecules are constructed. We see some dark stars. These are stars that have exhausted their supplies of nuclear fuel. It is very likely that some of the dark matter, especially in galaxy haloes, is made up of conventional 'baryonic' matter, the stuff that stars are made of. However there is so much dark matter, some 90 per cent of the mass of the universe, that we can be reasonably confident that we need a more exotic explanation for the dark matter. Most of the dark matter almost certainly consists of hitherto undiscovered, weakly interacting elementary particles.

Ordinary dark matter

The only halo dark matter detected to date is baryonic. The favoured baryonic dark matter candidates are star-like objects, dubbed MACHOs, standing for massive compact halo objects. They have been discovered via two remarkable experiments.

One technique used is gravitational microlensing. The MACHOs orbit in the halo of our galaxy. If a MACHO passes very close to the line of sight towards a distant star, the otherwise invisible MACHO acts as a gravitational lens that temporarily magnifies the starlight as the MACHO travels around the Milky Way halo. The duration of such a brightening is a few weeks for a solar mass MACHO, but the microlensing events are very rare. Only about one background star in several million will be microlensed at a given time. The result of monitoring several million stars in a neighbouring galaxy, the Large Magellanic Cloud, over six years revealed about 20 events that displayed the characteristic rise and fall signature expected for microlensing. The duration of the microlensing event measures the mass of the MACHO. The number of events tells us the fraction of the dark matter in MACHOs. The events detected amount to at most 20 per cent of the mass of the dark halo within about 50 kiloparsecs of the Sun. The event durations suggest a characteristic MACHO mass of around 40 per cent of the mass of the Sun. It is possible that perverse types of rare variable stars could be mistaken for MACHOs, or dim stars in the Large Magellanic Cloud itself could be the lenses. But the known data do not favour either loophole.

An independent experiment found the presence of old white dwarfs in the halo stellar population. Perhaps these are good candidates for MACHOs. White dwarfs are burnt-out relics of stars like the Sun. They are stars that have exhausted their nuclear fuel, and just glow from their residual thermal energy. White dwarfs in the halo have been cooling down for ten billion years or more, and so are exceedingly dim, radiating at less than a hundredth of a per cent of a solar luminosity. The sun is destined to evolve into a white dwarf in about five billion years from now.

Nearby disc stars are slow moving, as they co-rotate around the galactic centre with the Sun. Halo stars are known to have high velocities, since they populate a spheroid and intersect the Milky Way disc, and in particular the solar neighbourhood, from all directions. Some relatively dim white dwarfs are known to be nearby halo objects because of their displacements in the sky, relative to the more distant fixed stars. So are there enough dim halo white dwarfs, with a unique high velocity signature, to account for the MACHOs?

Old white dwarfs can only be detected if they are not too distant, in practice meaning within a few hundred light years from the Sun. Nevertheless, this volume provides a fair sample of the local halo. Astronomers have found, using several years of data to study the angular displacements of the dim nearby stars relative to the distant 'fixed' stars, that the white dwarfs amount to at most 5 per cent of the halo mass. This means that old white dwarf stars most probably are not the MACHOs. Something more exotic is required.

Exotic dark matter

If the MACHO results are accepted, one needs a population of solar mass compact objects in the halo, but even these are not the dominant contributor to the halo dark matter. Most of the halo dark matter remains to be identified. This might consist of ordinary matter, so-called baryons, as long as the matter is packaged in some suitably exotic form, not detected by microlensing. One might imagine a vast population of interstellar comets and asteroids. Any objects that weighed less than about 1 per cent of the mass of the Earth would not have been picked up by gravitational microlensing experiments. The effects would simply have been too small. However, such objects would have to consist primarily of hydrogen, since stars could not have existed in sufficient numbers to produce enough heavy elements to constitute the dark matter. Hydrogen 'snowballs' would evaporate in the depths of interstellar space, since for the hydrogen to be in a solid state, its temperature would necessarily be below 3 kelvin.

There remains the possibility of new states of baryonic matter, such as nuggets of so-called strange matter, a form of quark matter. Quarks are the ultimate building blocks of the protons and neutrons that constitute baryonic matter. Under certain extreme conditions at very high energy, quarks can aggregate into strange matter that could, in principle, have condensed within the early moments of the Big Bang when quarks were the dominant form of particles. However, such speculations have not received any support from, for example, our attempts to reproduce a plasma of quarks in nuclear accelerators and thereby study experimentally the possible states of matter in the very early universe.

What is far more likely, however, is that most of the halo dark matter consists of particles that are quite distinct in their interactions from baryons, and in particular interact very weakly with the baryons that make up the luminous matter of the universe. These particles are fondly known as WIMPs, standing for weakly interacting massive particles. Theoretical particle physics provides several candidates for particles that might comprise such non-baryonic dark matter. Indeed, novel types of very weakly interacting elementary particles arise naturally within conventional extensions of what particle physicists hopefully call the Standard Model. This accounts for all of the known properties of the elementary particles.

The Standard Model of elementary particles has performed remarkably well in predicting the properties of quarks. According to the Standard Model, there are fundamentally distinct types of particles that come in three families. Each family has four members. The lightest particles in each family are the leptons – the electron, muon, and tau. Each family also contains two types of quarks. These baryons are heavier than their leptonic brothers, and are the building blocks of protons and neutrons. Finally, there is the neutrino, another fundamental particle that comes in three varieties as a partner to the electron, muon, and tau.

All the atoms we see in the heavens and on the Earth are made up of electrons and quarks. But the other fundamental particles are crucial to account for the properties of atomic nuclei and their interactions, and for the force that stops nuclei from flying apart. This is all in the Standard Model. However, there are a few considerations that have no explanation within the Standard Model. Perhaps the most striking is the origin of the differences in mass between electrons and quarks. Why are leptons light, why are baryons heavy?

Physicists are straining at the leash to go beyond the Standard Model of particle physics. But the data are, so far, exceedingly sparse, and give little guidance. One of the main aims of a giant particle-accelerating machine, such as the Large Hadron Collider under construction at CERN, will be to explore physics beyond the Standard Model. In the absence of data, one has to be guided by theory.

One simple idea is that of perfection at the beginning of the universe, a sort of paradise where all are equal. For every particle there was an anti-particle. The properties of matter were once symmetric. Today they decidedly are not. However, at sufficiently high density and temperature, all details of individual particles were obliterated. Physicists go even further. They believe that Nature itself should be fundamentally symmetric. This seems a very Platonic approach, but it has reaped dividends. Symmetry is a form of beauty, and perhaps Keats was correct in declaring 'Beauty is truth, truth beauty'.

Symmetry is a compelling idea that gave birth to the Standard Model of particle physics. The most natural extension of the Standard Model is an attempt to incorporate more complete ideas about symmetry into an improved model. The new theory is known as supersymmetry and postulates the existence of a host of massive weakly interacting massive particles that partner each of the known particles. These so-called neutralinos are our best candidate for the WIMP. The lightest of these is thought to be long-lived and should be left over from the Big Bang in sufficient numbers to have present-day densities comparable to or even exceeding that of ordinary matter. Future accelerator experiments are capable of providing indications that these relics of supersymmetry do exist. For the moment, WIMPs represent a glimmer in the theorist's eye.

At sufficiently high temperature, particles, along with their partner antiparticles, are created and destroyed. Energetic photons produce particle pairs, and the pairs annihilate back into photons. Soon after the Big Bang, WIMPS would have been prolific. As the universe expanded and cooled, it became energetically unfavourable for any new WIMP-like particles to be created. There was a dramatic reduction in the particle content of the universe when the temperature was too low for WIMPS to be produced. In fact, the fate of most WIMPs was to annihilate as they found a partner WIMP. Only a few WIMPS remain behind, to be the dark matter, because they failed to annihilate with one another as the density and temperature dropped.

The leftover WIMPs are, then, a relic from the past. Their abundance is determined by how many survived, and this depends on their interaction strength, which is unknown from fundamental particle physics. A sufficient number of stable WIMPs, most probably the neutralino survivors of supersymmetry theory, may have survived to make up the dark matter. Their masses are large compared to protons, so although few survive, the survivors greatly outweigh protons in contributing to the mass density of the universe. Just occasionally, neutralinos still annihilate each other in the haloes of galaxies. Observable debris may result. Dark matter is not totally dark.

Neutralinos are everywhere. In particular, the halo of the Milky Way has about 100 per cubic metre. About one million pass through our bodies every second. But it is not easy to trap one, since their interactions are so weak. Indeed, if the neutralinos were strongly interacting particles, far fewer would have survived.

Even so, sensitive experiments have been devised that are capable of detecting the neutralinos that are passing through our laboratories. The idea is to go deep underground, ideally in a deep mine, where one is effectively camouflaged from the cosmic rays. Cosmic rays are protons or heavier nuclei. They are strongly interacting, and so are blocked by thick rock from entering the deep underground cavern where the experiment is being run.

One such experiment is carried out a kilometre underground in the Boulby potash mine in Yorkshire. Another is in a laboratory hacked out of a service road off the Mont Blanc tunnel, under an equivalent amount of rock. These experiments are searching for the modulation of the cosmic flux of neutralinos by the Earth's motion. The effect is similar to the experience of driving a car in the rain. The rain appears to fall at an angle that depends on the speed and direction of the car. The Earth's motion is only 10 per cent of the typical velocity of the WIMPs, and so the effect is tiny, amounting to only one part in ten thousand. Every year, the signal should peak and six months later it should reach an annual minimum. If such a modulation is detected, we can be sure that the neutralinos were extrasolar in origin.

One experiment claims to have detected just this effect. A vast vat containing, in its most recent incarnation, a quarter of a tonne of liquid scintillator, a dilute solution of sodium iodide, and situated in the heart of the Gran Sasso underground laboratory in the Abruzzo mountains in Italy, has been collecting data for seven years. The scientists, led by Rita Bernabei of the University of Rome 'Tor Vergata', provide evidence for the sought-after annual modulation that is caused by the passage of a wind of WIMPs, due to the Earth's motion. The heavy WIMP particles impact on sodium atoms, whose recoil energies generate tiny light flashes in the scintillator fluid. However, other experiments, which claim higher sensitivity to WIMPs, including collaborations in the USA, France, and the UK, have failed to confirm the result of the Italian group. A definitive result must await future and still more sensitive experiments.

Of course, there is no guarantee that neutralinos exist. Other dark matter particle candidates have masses that could be anything from a trillionth of a proton mass to a trillion or more proton masses. Yet the direct detection experiments are sensitive only to a narrow range of neutralino masses, between a proton mass and a few hundred proton masses. The lower end of this range is excluded by accelerator experiments. We know that neutralinos must weigh at least fifty proton masses. So the accessible range is very limited. It reminds one of the drunk who, losing his car keys, returns to search under the nearest lamppost.

Most people would not lay a bet against odds of this sort. Yet this has not dissuaded the proponents of neutralino detection technology from looking, possibly for nothing. One has to be exceedingly patient. It may take weeks or months for a single neutralino to collide with the detector material. Most would pass straight through the detector and, for that matter, straight through the Earth.

Imagine the excitement among the assembled scientists when a detection is registered. Vast numbers of simulations will have been carried out, to weed out the false alarms. But the elusive neutralino signal could nevertheless easily be confused with extraneous signals. Even if one has eliminated the sources of irrelevant events, such as cosmic rays, there are some that cannot be avoided. Potassium, a common element found everywhere in rock, has an isotope that is radioactively unstable. Deep in the total darkness under a kilometre or more of rock, beneath a mountain, or deep underground, the feeble glimmer of decaying potassium is worrisome for the experimentalist searching for a signal that is even fainter.

A more mundane example is radon emission from the rock. This and similar low-level radioactivity is endemic to rocks, and even to bricks. Houses made of brick have measurable radon contamination. This would contribute to the overall background against which any WIMP signal must be sought. It is even possible that there might be seasonal variations, induced by temperature changes that could mimic the desired modulation signal. Most terrestrial signals would not show such an annual modulation, however.

It will be necessary to dissect the signal in detail, to develop detection techniques that are sensitive to the energy and direction of the exceedingly rare interactions of a massive neutralino that collides with cryogenically cooled atoms of germanium crystals, or something equally sensitive, in the detector. The annual variation offers the best prospect for distinguishing a WIMP. But the search will be long and difficult, and is likely to have many false alarms. Nevertheless, researchers are persevering, constructing ever more sophisticated detectors in remote underground sites.

A typical neutralino might weigh 100 proton masses. Imagine one neutralino encountering another neutralino in the halo of the Milky Way. The encounter must be self-destructive. It is like the annihilation event when a proton meets an antiproton, but the neutralino event is much more violent because so much more mass is liberated into energy. The annihilation products include very energetic particles, such as pairs of protons and antiprotons, electrons and positrons, as well as gamma rays. A diffuse glow of gamma rays is predicted in the halo if neutralinos are the predominant dark matter component, and that gives another way of detection. Gamma ray and cosmic ray telescopes are being designed to search for such signals above the Earth's atmosphere. However, cosmic ray protons interacting with heavy interstellar atoms also generate secondary antiprotons and positrons. A way is needed to disentangle the primary annihilation signature from the secondary signal. This can be accomplished with detailed study of the energy distribution of the cosmic rays.

Another possible signature is from the neutrinos with high energies that are also produced in annihilations. Neutrinos rarely interact with matter, but when they do, brief flashes of light are generated. These are very weak, and one needs to monitor a large volume of matter to find a signal. Such light flashes are seen in the atmosphere, but are primarily induced by cosmic rays. Neutrinos can pass through the Earth, and give a unique upwards signal if the experiment can detect direction. Experiments that monitor vast amounts of water or ice are under development for high-energy neutrino searches. The ANTARES experiment, off the coast of Toulon, monitors about a tenth of a square kilometre of the Mediterranean, with ten strings of photomultipliers (very sensitive detectors) reaching to a depth of a third of a kilometre. The IceCube project, at the South Pole, has strings of photomultipliers embedded in the ice to a depth of 1.5 kilometres, where it is completely dark, and monitors a cubic kilometre of ice. The direction of signals is given by viewing the sequence of light flashes as detectors are triggered on adjacent strings.

These are potential indirect detections of neutralinos that annihilate in either the Earth or the Sun, and generate high-energy neutrinos. Direct detection is also feasible, as we have seen, in sensitive cryogenic laboratory detectors that are installed deep underground to avoid contamination by cosmic ray induced signals. However, it will take another decade before the experiments now being designed attain sufficient sensitivity to make definitive exploration of neutralinos feasible via direct detection. It may be that the indirect signal will provide the critical evidence in the race for identification of dark matter.

The first hints of the existence of neutralinos, however, may come from elsewhere. Over a similar timescale, particle accelerators, most notably the Large Hadron Collider at CERN, scheduled to be operative in 2007, will be capable of searching for evidence that neutralinos exist. The high-energy collisions in these machines produce jets of energetic particles and anti-particles that are ejected during the collision. The jets are splattered in opposite directions, at right angles to the collision direction. While any weakly interacting WIMP would be invisible, it carries off momentum that must be balanced by a detectable jet of particles. A one-sided jet would be evidence for a supersymmetric particle, and hence for the neutralino or WIMP hypothesis.

Actually even the Large Hadron Collider may not be sensitive enough. The problem is that smashing hadrons together results in extremely messy debris, since the hadrons are not fundamental particles but are composites. It may prove impossible to extract the elusive neutralino signal. Instead, the answer will have to await the construction of a future Linear Collider, a device that accelerates electrons and positrons to ultrahigh energies. Such a machine, which could not be built before 2020 at a projected cost of at least 15 billion dollars, could provide a sufficiently clean signal of high energy interactions that supersymmetry would be detectable.

Yet another direction for studying non-baryonic dark matter has come from cosmology, and in particular from the theory of how the large-scale structure of the universe must have formed. Simulations of the formation of large-scale cosmic structures provide persuasive evidence that dark matter is made up of very weakly interacting particles that would have started out much colder than the baryons. The weakness of their interactions means that they were unlocked from the grip of the radiation fireball in the Big Bang much earlier than were ordinary particles, and so they are much colder. These particles constitute 'cold dark matter'. Such cold matter is much more unstable to gravitational condensation than ordinary matter, which was relatively hot in the early universe. Pressure forces stabilize the matter against gravitational condensation, and only cold matter can aggregate freely. Since the cold dark matter is the dominant form of matter, it controls the local gravity field, and hence the formation of structure by gravitational instability. Supersymmetric particles could well provide the 'cold dark matter' that cosmology requires.

6 The Invisible Cosmos

Strange to say, the luminous world is the invisible world, the luminous world is that which we do not see. Our eyes of flesh see only night.

Victor Hugo

Sit down before fact as a little child, be prepared to give up every preconceived notion, follow humbly wherever and to whatever abyss nature leads, or you shall learn nothing.

Thomas Huxley

Astronomers used to be extraordinarily biased. Our eyes and telescopes, at least until a half-century ago, were limited to optical wavelengths. In fact, the visible part of the spectrum, spanning the colours of the rainbow, violet to red, is only a tiny part of the electromagnetic spectrum. Light invisible to the human eye is produced by cosmic sources over a vast span of wavelengths, from tens of metres to trillionths of a metre, from radio waves to gamma rays. An X-ray survey of the sky reveals a totally different vista from that visible to the optical astronomer on a dark night. An infrared view is completely different from both.

The X-ray universe

View the universe through an X-ray telescope in space and the perspectives differ dramatically from what the Earth-bound observer can see. The X-rays from cosmic sources, fortunately for us, are blocked by the Earth's atmosphere. Even the Sun, especially during its most active phases, is a source of X-rays that, were it not for our atmospheric shield, would have deleterious consequences for life on Earth.

Much of what we know about neutron stars and black holes has come from studying the universe at X-ray wavelengths. We have deciphered the masses of neutron stars. The energies of photons are measured in electronvolts or kiloelectronvolts. Visible light has photon energies of a few electronvolts, X-radiation of a few kiloelectronvolts, and gamma rays of a few megaelectronvolts.

We study neutron stars and black holes at kiloelectronvolt energies using telescopes on board space satellites. The observations are performed with experiments such as NASA's CHANDRA X-ray Observatory and ESA's NEWTON X-ray Telescope. Both provide images and spectra of objects in the X-ray sky. Most of the neutron stars are in our galaxy.

Further afield, the high-energy view of the universe reveals it to contain cauldrons of hot gas. The largest of these are the great galaxy clusters. Galaxy clusters are diffuse souces of X-ray emission. These are the reservoirs of the gas from which the galaxies formed. In fact, the X-ray sky is dominated by many point-like sources. In the nearby universe, these typically are binary stars, in which a compact star, either a neutron star or a black hole, is accreting prolific amounts of gas from its companion star.

Active galactic nuclei are the most common X-ray sources in the distant universe. The universe was once a hive of activity. As we look back in time, to when the universe was a third or less of its present age, we see many galaxies with active nuclei. In fact, some nuclei are so bright that they are all we can see. It is as though one is blinded at night by a bright source of light. We now understand how the central engine of an active galactic nucleus in reality is a supermassive black hole that is fed by accretion of gas from its environment.

The infrared universe

When one surveys the universe in the far infrared, one sees primarily the emission from the dust. In regions of star birth, dust accumulates at high enough density and opacity to shroud the birth events completely at optical wavelengths. Ordinary light is absorbed and scattered by the dust. The dust re-emits the absorbed light in the infrared part of the spectrum. There are also very old, cold stars that glow feebly in the infrared. But mostly one is viewing the young side of the universe.

The universe is a dusty place. Our Milky Way is a prime example, indeed our solar system is another. Interstellar dust begins its life cycle in the ejecta from evolved stars, called red giants. As stars age, thermonuclear reactions proceed at ever-increasing temperature in the stellar core. Carbon and other heavier elements are synthesized, until the star exhausts its supply of nuclear fuel. The effect of the enormous central heat is to swell up the atmosphere of the star. The star becomes a red giant. The outer layers of the star are eventually expelled. A white dwarf star is created at the centre. The expelled envelope forms what is called a planetary nebula. William Herschel, who, observing in the early days of telescopes, could only note a vaguely planet-like appearance, coined this phrase.

The ejected gas layers cool and the non-volatile heavy elements condense into tiny solid particles, of graphite and quartz-like materials. These constitute dust grains, or at least the refractory cores of grains, that are found in space. These grains permeate interstellar space, and are occasionally captured by the upper atmosphere of the Earth. The small grains end up in interstellar clouds of gas, which cool to such low temperatures that even the more volatile and abundant elements condense out on the dust grains. The refractory cores are covered with mantles of ices. About 1 per cent of the mass of the interstellar medium is in tiny solid particles, about a thousandth of the diameter of a grain of sand.

The space between the stars contains clouds of diffuse gas and dust. Most of the interstellar matter is in clouds, although there is a pervasive interstellar medium between the clouds. Astronomers detect cloud-like aggregates of gas and dust, because the dust in the clouds both scatters and absorbs incident starlight. Shorter wavelengths are scattered more, so the redder light remains. Scattering reddens the light of background stars.

One consequence of scattering is that some interstellar clouds are so dense that they are completely opaque to starlight. The dust effectively scatters and absorbs all the optical radiation. However, the light absorbed by the dust is reradiated in the infrared, and there is a general infrared glow from dusty clouds. Stars form in nebulae that are permeated by dust. We cannot view star formation in the optical spectrum. Because of the rotation of the nebula, the dust aggregates into a disc that surrounds the central forming star.

The disc is cold and dense. Dust settles into the mid-plane, much as sand settles in the aftermath of a sand storm. When the solar system formed, a belt of dust had accumulated around the forming sun. Larger and larger dust grains coalesced. Eventually, rocks the size of tiny asteroids formed. These are dubbed planetesimals, and are the hypothesized missing link between the dust grains and typical asteroids or planets. Planetesimals are the building blocks of the solar system, and vestigial relics may remain today in the asteroid belts.

Planets eventually coalesce out of the dust. Traces of the dust are left in the ecliptic plane (the plane on which the orbits of the planets lie), and on a dark night we see this dust reflecting sunlight as the dim glow of the zodiacal light. But to an infrared observer this would be a dazzling belt of light that obscured up to a quarter of the sky. The local dust is warmed by the sun, and emits in the near infrared. In the far infrared, the local dust is invisible and one can peer far away.

In the Milky Way, the accumulation of dust in the galactic plane prevents us from seeing in optical light more than a thousand light years away. This is a fraction of the distance to the galactic centre. Optical light is scattered by dust. In the infrared, all is revealed. We can see the stars that orbit the very nucleus of our galaxy, where a supermassive black hole is lurking. Indeed, we infer the existence of this invisible monster by studying the orbits of the central stars, at infrared wavelengths.

In the distant universe, forming stars are shrouded by dust. We can only penetrate the murky layers of dust at infrared wavelengths. Consequently, we need to view the universe in the infrared in order to be able to study how galaxies formed. At least half of all starlight is absorbed and re-emitted by dust. Infrared astronomy plays an essential role in providing a complete census of distant star formation in the universe. But there is still more to be seen, especially when stars die. Violent phenomena such as explosive star deaths are associated with emission of radio waves and gamma rays. The precursors, however, are seen in the ultraviolet.

The ultraviolet universe

Nearby star formation can be probed at ultraviolet wavelengths, where very hot, massive stars emit most of their radiation. The light from star-forming galaxies is dominated by the contributions from the massive stars and the nebulae that they excite. The most abundant atoms in the universe, hydrogen and helium, have their strongest spectral features in the ultraviolet. So do other abundant elements, such as carbon and oxygen. An electron is excited by a collision and jumps into a higher energy state. This process results in absorption of a quantum of light, and is inevitably followed by emission of a photon as the electron jumps back down.

Emission by an electron dropping to its lowest energy state is usually in the ultraviolet. Most of the cold gas in the universe is sitting in its lowest energy state. When a photon incident on a hydrogen atom causes absorption by inducing an electron to jump to a higher state, the absorption features that result are in the ultraviolet.

Fortunately for ground-based observers, there are intermediate steps in the cascade of electrons from higher to lower energy states that result in optical lines. However, for absorption by the most abundant species, this is usually not the case. Only for relatively rare elements, such as sodium and potassium, does the absorption occur at visible wavelengths. These elements are nevertheless found to be common throughout interstellar space and to demarcate interstellar clouds. The most common element, hydrogen, is far more elusive.

Hydrogen absorption lines are studied by telescopes in space for the nearby interstellar and intergalactic gas. However, absorption by the remote intergalactic medium is redshifted from the ultraviolet into the optical part of the spectrum. We learn that at high redshift the universe has large amounts of intergalactic gas, at least twice as much as is seen locally. Some early massive star formation would have occurred very early, perhaps at a redshift of 10, enough to pollute the intergalactic gas before the large galaxies formed. An ultraviolet view of the universe contains other surprises. Newly formed white dwarfs are very hot. While dim in the optical, they are very prominent in the ultraviolet. Galaxies that are forming stars have rather regular spiral patterns in the optical. An ultraviolet view is quite different, however. Dust obscuration plays a major role. Bright knots dominate where massive stars are forming, and the interstellar medium is impressively chaotic in appearance, full of bubbles and filaments. Most of this emission is due to gas at hundreds of thousands of degrees. The optical radiation from gas this hot is negligible compared to the ultraviolet emission.

The radio universe

Dust is no barrier for radio waves. The radio universe was discovered only in the mid-twentieth century. The Milky Way was found to be a source of radio noise. The sun emits occasional bursts of radio noise. And as the technology improved, astronomers found that the entire universe was bathed in a sea of radio waves. The sources of the radio waves ranged from the remnants of ancient supernova explosions, to the nebulae surrounding massive young stars in our Milky Way, to remote galaxies. Normal galaxies like our own were found to be relatively weak emitters, but galaxies in an apparent state of merging were among the brightest of the extragalactic radio sources.

New phenomena were discovered as the radio studies progressed. One of the best known is a pulsar. A pulsar is a rapidly spinning, highly magnetized neutron star that emits intense beams of radio emission, intersecting the terrestrial observer rather as would the beam of a cosmic lighthouse. Thousands of pulsars have been discovered in the Milky Way, with rotation periods that range from seconds to milliseconds. The neutron stars formed out of the explosions of massive stars, and one such explosion was recent enough to be in the historical records. The Crab supernova event was observed on 4 July 1054, when Chinese astronomers noted a 'guest' star in the constellation of Taurus, visible for about two weeks. Modern observations of the Crab Nebula reveal that the nebulosity is expanding at thousands of kilometres per second, consistent with its explosive origin in 1054. The neutron star has been identified, first as a radio pulsar, and then as an optical pulsating source. The Crab pulsar is also detected in X-rays and in gamma rays, emitting pulses of radiation at the characteristic frequency of 30 milliseconds.

The first example of a pair of colliding galaxies came from the optical follow-up of one of the brightest radio sources in the sky, in Centaurus. Radio astronomy has led to the discovery of new populations of galaxies, radio galaxies, characterized by radio emission that is thousands of times stronger than that from our own galaxy. Astronomers believe that the radio outburst is triggered by the fuelling of a supermassive black hole due to the accumulation of gas during the mergers of pairs of interacting or colliding galaxies.

The high energy universe

At gamma ray frequencies, one sees the spectacular events that power the most luminous objects in the universe. These are truly cosmic lighthouses, sources that are highly beamed and compact, yet immensely more luminous than any other steadily shining objects. These are active galactic nuclei and quasars, to be described below. Their relics are found in the nuclei of nearly all nearby galaxies. What appear to be monster black holes are found. Of course, one does not see the black holes directly or even, for most of these nearby systems, indirectly. Their presence is inferred from the requirement that a large amount of mass must be concentrated in an exceedingly small volume.

However, even more energetic phenomena have been discovered. The brightest objects in the universe are very short-lived. They are transitory, only shining for a few seconds, but are far more luminous in gamma rays than even the most luminous galaxies. Gamma ray burst sources produce prodigious amounts of the most energetic photons. Gamma rays have energies of millions of electronvolts, or a million times that of ordinary light, and a thousand times more than X-rays. About once every few seconds a star explodes as a supernova somewhere in the universe. One exploding star in a thousand explodes with far more vigour than the typical supernova. We call these hypernovae. Hypernovae were massive stars, weighing at least 25 suns, that collapsed. The core forms a black hole, and vast amounts of energy are released that eject the outer layers of the star. A hypernova releases up to a hundred times more kinetic energy than a typical supernova. In some cases, much of the explosion energy comes out as a violent burst of gamma rays. It is as though a fraction of the rest mass of a star were almost instantaneously to vaporize into high energy radiation.

The gamma ray burst is over after a few seconds, but it is followed by a more long-lived faint optical afterglow. The luminosity of the burst is prodigious. It is a trillion times brighter than a supernova, which itself is as bright as the entire Milky Way, and is observable in gamma rays. Satellites launched by the USA and USSR in the 1960s to search for evidence of clandestine testing of thermonuclear weapons in space discovered the cosmic gamma ray bursts. Modern satellites detect about one a day. It is most likely that the gamma rays are highly beamed, so that the true frequency is hundreds of times larger. The gamma ray spectrum shows the characteristic imprint of enriched ejecta from a massive star outflowing at a tenth of the speed of light. One can detect gamma ray bursts at the furthest ends of the universe.

The optical afterglow persists for a few weeks after the explosion. In one case when the afterglow was detected almost simultaneously with the gamma ray burst, it reached a brightness of a star only ten times dimmer than the faintest naked eye stars before fading away. Yet the gamma ray burst was at a redshift beyond 1.6, where entire galaxies have the equivalent brightness of stars that are a million times fainter. Other bursts have been detected at redshifts beyond 4. There is every prospect that with an X-ray satellite launched in 2004 called SWIFT that is aimed at detecting bursts, we will be able to find them when the first stars were forming, at redshifts of 10 or larger.

Other denizens of the high energy universe are the particles from space that we call cosmic rays. These are charged particles, typically protons and electrons, that have been accelerated to enormous energies. Cosmic rays are produced in explosions of stars and further accelerated in interstellar gas shocks to high energies. The remnants of exploding stars include the central pulsars, which are rapidly rotating highly magnetized neutron stars and consequently excellent accelerators of charged particles. Pulsars are injectors of high energy particles. Further acceleration occurs in supernova remnants, the ejected debris from exploding stars that shock and sweep up interstellar gas. Shocks are an ideal environment for particle acceleration.

Wilbur Hess, an American who flew unexposed photographic plates in high altitude balloons in the second decade of the twentieth century, discovered cosmic rays. To his amazement, when he developed the photographic plates in his dark room, he found that the unexposed emulsions were streaked. The streaks turned out to be tracks etched by high-energy particles from space that ionized the silver emulsion. Later experiments showed that the Earth is bombarded by cosmic rays, most of which, fortunately for us, do not penetrate the lower atmosphere. If they did, the number of genetic mutations and cancers would surely rise markedly. Cosmic rays provided a fertile hunting ground for particle physics at a time when powerful terrestrial particle accelerators were only just beginning to be constructed. Indeed, a new elementary particle, the muon, which is intermediate in mass between the electron and the proton, was first discovered in the cosmic rays. It is a secondary particle, normally very short-lived, and produced in the course of collisions of cosmic rays with atmospheric molecules. Only much later were muons discovered in particle accelerators, where beams of high energy particles were collided and the debris was studied.

We see muons in the cosmic rays because of the phenomenon of time dilation. Produced high in the atmosphere, the muons take milliseconds to reach the earth. They should have decayed: in its rest frame a muon only survives a microsecond before spontaneously decaying. However, the cosmic ray muons have such high energies that, as predicted by Einstein's theory of relativity, time is slowed down in the observer's frame. One views a high energy muon before it has had time to decay spontaneously, as it would do were it at rest, in only a millionth of a second.

Heavy elements are also present in enhanced amounts in the cosmic rays. This tells us that the acceleration occurred in the vicinities of exploding stars, perhaps in the explosion itself. The most energetic of the cosmic ray particles have energies of a billion ergs. This means that a single cosmic ray proton has the same energy as that of a rock weighing a kilogram dropped from the top of the Eiffel Tower. As we have seen, the atmosphere shields us from direct cosmic ray impacts. Moreover, these very energetic cosmic rays are rather rare. Only about one per hundred square kilometres per year at the very highest energies impacts the Earth. Each impact at the top of the atmosphere produces a shower of energetic particles, including muons that can penetrate through the Earth's atmosphere.

Experiments to detect the ultra-high energy cosmic rays look for muons produced in the air showers. They also search for light flashes produced as the cosmic rays interact and are slowed down in the Earth's atmosphere. It is necessary for the detectors to cover hundreds of square kilometres in order to have a reasonable chance of detecting the most energetic cosmic rays. One such experiment, the Pierre Auger Observatory, consists of an array of 1600 detectors spaced 1.5 kilometres apart and covering an area of 3000 square kilometres in Mendoza, Argentina. The detectors are 10 square metre surface area tanks of purified water, lined with a highly reflecting surface. They are designed to look for light flashes produced when particles travel faster than light through the atmosphere or water. (It is only in a vacuum that the speed of light attains its maximum value, and one that no particle can exceed. In water or even in air, the speed of light is reduced.) Energetic particles such as high energy muons produced in the course of the interactions of cosmic rays with the atmosphere are therefore travelling at superlight speed. This has the effect of stimulating the water or air atoms to emit a faint bluish flash of radiation, known as Cerenkov light, rather analogous to the sonic boom of a supersonic aircraft. The effect is more concentrated in water, and only lasts for a nanosecond. Muons from the atmospheric air shower hit the water and emit a flash of Cerenkov light that is monitored by photocells. The Global Positioning Network of satellites is used to provide accurate timing for each of the detectors so that the path of the incoming cosmic ray at the top of the Earth's atmosphere can be reconstructed to an angular precision of a third of a degree.

The ultra-high energy cosmic rays are accelerated in shocks in the intergalactic medium, produced where the energetic outflows from radio galaxies decelerate by running into the ambient gas. They are also accelerated in the vicinities of the supermassive black holes in the nuclei of active galaxies. The most energetic cosmic rays bear witness to exotic physics that occurs in the nuclei of the most active galaxies. Conditions are surely extreme in the vicinities of supermassive black holes. It is likely that only near very massive black holes are the conditions sufficiently favourable to be able to accelerate the cosmic rays to the highest energies that are observed.

7 Supermassive Black Holes and Galaxy Formation

... time is the longest distance between two places.

Tennessee Williams

I have been a stranger in a strange land.

Exodus

The black holes of nature are the most perfect microcosmic objects there are in the universe ... they are the simplest objects as well.

Subrahmanyan Chandrasekhar

Predicted by the theory of general relativity, black holes are among the strangest objects known to exist. A point in space usually has a past and a future. An explosion is an example of a point in space–time. The explosion is both at a specific point in space and at a specific time. Any observer, or any light signal, can be said to move towards the future. If it travels far enough and long enough, it is in theory eventually able to reach any other point in space at some future time. The one exception to this rule is experienced by an object near a black hole. A black hole may be defined to be a region in space–time where there are events from which there is no escape, even for light signals. A black hole is a trapped surface: cross this surface, and there is no going back!

The size of a black hole

The radius of a black hole, named after the astronomer Karl Schwarzschild, is proportional to the mass of the hole. For a black hole of one solar mass, one finds that the Schwarzschild radius is only three kilometres. For supermassive black holes, which are thought to contain up to one billion solar masses, the Schwarzschild radius would only be the diameter of the Earth's orbit around the Sun, exceedingly small by astronomical standards. Planets would not be sucked in by a black hole the mass of the Sun. The Earth would just as happily orbit if the Sun were replaced with a black hole of the same mass, since the gravity field at this distance would be unchanged. A rocket ship visiting a black hole would feel that gravity was no different from that around a star, until the spacecraft approached within a few Schwarzschild radii of the hole. Then tidal forces would begin to affect the spacecraft in a very dangerous way.

Seeking a black hole

Black holes are detected by their gravitational effects on neighbouring stars. Stellar mass black holes are the end-states of very massive stars. Stars of initial mass greater than about 25 solar masses are destined to end up as black holes. There is no known pressure that can stop the collapse of a stellar core if its mass is more than a few solar masses. Not even a pressure as extreme as the quantum pressure of overlapping atoms, or the quantum pressure of overlapping neutrons, can help to avert the final collapse. About 30 such black holes have been discovered in our galaxy and in nearby galaxies as members of close binary systems. The black hole is close enough to its companion star so that when the star evolves to become a giant or a supergiant, matter from the star's atmosphere spews on to the black hole. The accreted gas spirals on to the black hole, forming a disc that slowly falls into the black hole itself. As the gas falls in, it heats up and emits X-rays. The orbital characteristics enable one to determine the mass of the X-ray emitting star and show that it must be a black hole.

Some galaxies have vast powerhouses of energy in their cores. Their spectra show strong emission lines characteristic of gas moving at a few per cent of the speed of light. Even brighter are the quasars. The spectra of these objects indicate gas turbulence velocities of up to 20 per cent of the speed of light. Quasars are now known to be the ultraluminous cores of otherwise normal galaxies.

The remarkable phenomena that can occur in the nuclei of galaxies

The energetic phenomena occurring in quasars and active galactic nuclei are thought to be powered by supermassive black holes. Gas shed by evolving stars in the inner galaxy is believed to be fuelling a central massive black hole. If matter falls towards the centre on a radial orbit, it is accelerated to near the speed of light before reaching the Schwarzschild radius. Because the galaxy is rotating, the gas has angular momentum, and for the most part it is expected to spiral in as it loses energy. The gas accretes into a dense disc that surrounds the central black hole. At this stage, the gas is hot enough to emit X-rays. The gas loses energy as it radiates and spirals closer until it disappears across the event horizon of the black hole, the point of no return.

Quasars are intense sources of X-rays, coming from gas that is near the black hole. Star collisions also play a role in determining the properties of the central black hole. Stars are closer together in the dense nuclei of galaxies, and the debris released in stellar collisions is likely to provide an additional source of fuel for the black hole.

Even a trickle of a hundredth of a solar mass per year, a tiny fraction of the several solar masses annually consumed in forming stars in a galaxy like the Milky Way, suffices to produce strong activity around the central black hole. About 1 per cent of all spiral galaxies are characterized by having extremely bright nuclei, or 'active galactic nuclei', powered by supermassive black holes. Such galaxies are known as Seyfert galaxies, named after the astronomer Carl Seyfert.

Quasars are far brighter objects, and their central supermassive black holes consume hundreds of solar masses per year in order to fuel the prodigious outpourings of energy. The emission from a quasar, equivalent to the light from a thousand or more Milky Way galaxies, is produced in a region that is only a few light-minutes across, less than the size of the solar system. We can infer the size because the quasar emission is observed to be variable over timescales of hours and even minutes. The region of emission must be highly compact if the light varies coherently over such short timescales.

The active phase must be short-lived, lasting perhaps 1 per cent of the present age of the universe, as otherwise far too much energy would be generated than can be accounted for. The power source of quasars is also conjectured to be accretion on to a supermassive black hole. Most galaxies may have undergone an active phase in their past, and should therefore possess massive black holes in their nuclei. Nowadays 99 per cent of galaxies have faint and unremarkable nuclei: their central supermassive black holes must be completely inert, with no matter falling into them.

Quasars are rare in the nearby universe, but were common long ago. Quasars have now been discovered at redshifts beyond 6, and are the most powerful direct probes that we have so far of the early universe. We find that the number of quasars rises rapidly as we look far into the past. The number is so large at an epoch when the universe was a quarter of its present size that we infer that every giant galaxy may once have harboured a quasar in its nucleus.

Now all of this is but a memory. The ambient conditions are far less extreme, and for whatever reason, the black hole is sitting quiescently at the centre of the galaxy, perhaps waiting for a fresh supply of fuel. Supermassive black holes are found in the cores of many nearby galaxies. All are generally passive today. One cannot tell they are there other than by the indirect influence on the orbits of nearby stars.

How to detect supermassive black holes

Confirmation has come from studies of the nuclei of nearby galaxies, which reveal the presence of immense matter concentrations that can only be supermassive black holes. The motions of the gas clouds or stars increase within the central light year. This indicates a vast mass concentration that cannot be due to, say, a compact cluster of stars: the stars would collide. The only interpretation that makes any sense is the presence of a central supermassive black hole. Perhaps the most remarkable result is that the mass of the spheroidal component of old stars is tightly correlated with the black hole mass.

The giant galaxy Messier 87 has a central black hole of three billion solar masses, while the Andromeda galaxy has a central black hole of about a million solar masses. In our own galactic centre, observations over a five-year baseline have reconstructed the three-dimensional motions of stars within a thousandth of a parsec of the centre. The stars are observed to be orbiting the centre of mass of the galaxy at ever greater velocities as one approaches the centre. The stellar orbits reveal the presence of a central black hole in our Milky Way galaxy of three million solar masses.

Only a supermassive black hole could be responsible for the central mass concentration. Yet this immense black hole, despite being surrounded by stars and gas clouds, is virtually invisible. Indeed, in nearly every nearby example, the supermassive black holes are quiescent and dim: only their gravitational footprint tells the tale.

The dimness of the nearby supermassive black holes is puzzling. One would have expected some evidence of emission of radiation produced by accreting gas, visible at X-ray frequencies. A black hole normally is expected to light up its environment. Gas accretes, heats up as it experiences the tug of the black hole's gravity, and is compressed. The gas should be radiating prolifically in X-rays. In the case of our galaxy and the Andromeda galaxy, the resulting X-radiation expected from infalling gas is simply not observed. At the current epoch in the universe, supermassive black holes are remarkably passive objects.

The scaling of black hole mass with old stellar content of the galaxy means that supermassive black hole formation is intimately related to the process of galaxy formation. The gas-rich environment of the forming galaxy provides the ideal conditions for stimulating black hole formation. What is not so clear is the effect that the supermassive black hole and the outflows that will inevitably be generated would have on the dark matter as well as on the process of galaxy formation. This remains a subject of intense speculation. It may be that the difficulties arising in cold dark matter modelling of galaxy formation can be resolved by properly incorporating black hole physics.

The violent universe

Let us try to piece together an outline of how both galaxies and supermassive black holes must have formed. Our current view of cosmology is that the universe has evolved from a highly homogeneous and isotropic state on all scales. It has evolved into one that is still statistically homogeneous and isotropic but characterized by inhomogeneity in the form of large-scale structure on scales of up to tens of megaparsecs.

Hierarchical structure formation is characterized by a sequence of mergers. Dwarf galaxy mass objects merge to form more and more massive systems. This culminates in cluster and supercluster formation. All of this applies to the dark matter haloes. Dissipationless merging and clustering is a characteristic of the dark matter. The cold dark matter clusters in a sort of hierarchy, forming ever more massive haloes with increasing epoch. The mass in baryons constitutes approximately 15 per cent of the mass in dark matter. The baryonic component is in the form of gas that dissipates strongly, forming cold cores that are embedded within massive haloes. Eventually the baryons fragment into stars. Collisions between haloes play a central role in galaxy formation theory. Galaxy formation is a violent process.

The most important issue in galaxy formation concerns the mechanism that regulates the rate and efficiency at which the gas is converted into stars. It is here that collisions between gas clouds play a central role. The stellar components of galaxies can be categorized into discs and spheroids, as well as an unclassifiable category, irregulars. It is the latter that often provide direct evidence for triggering of star formation by tidal interactions and mergers of galaxies. Mild tidal interactions drive clouds to cluster together. However, the violence of a merger has a more dramatic outcome. Gas clouds collide, undergo shock waves, and are compressed, resulting in their fragmentation into stars.

After a few orbital timescales, large-scale regularity tends to prevail and either discs or spheroids of stars form. Discs are undergoing continuous star formation, whereas spheroids formed their stars long ago and so are predominantly old. It would seem that there are two distinct modes of star formation. Discs require a gas supply in order to maintain a low and relatively inefficient but long-lived rate of star formation. Discs form in low-density environments in which major mergers are rare. Nevertheless, minor mergers certainly occurred. Spheroids, on the other hand, formed in a burst, and star formation ceased after hundreds of millions of years had elapsed. Tidal interactions and mergers control the history of star formation in galaxies. Ellipticals formed via major mergers in denser environments.

Major mergers drive the gas into a compact core a few hundred parsecs across. Here the ensuing burst of star formation is initially completely shrouded by dust. Minor mergers and collisions result in tidal shocks and enhanced star formation. This has the effect of driving gas inwards to form central bulges.

This modern, collision-driven scenario of galaxy formation is indicated by several kinds of observations. Irregular galaxies constitute about 1 per cent of all galaxies at present but become the dominant component at high redshift. Ultraluminous infrared galaxies are rare today and were far more common in the past. They inevitably show evidence for a recent or an ongoing merger. The inferred star formation rate in an ultraluminous infrared galaxy is similar to that expected for a spheroid in formation. Nearby examples of starbursts show a light profile in the near infrared characteristic of a spheroid.

The diffuse far infrared background contains a comparable energy density to the optical background radiation, and as it was generated earlier, it would seem that most of the star formation in the universe was probably shrouded by dust. The dust clears as a consequence of 'superwinds' driven by many supernovae associated with the deaths of massive stars produced in the starburst. These winds enrich the intracluster medium, and metal-poor spheroids of stars remain behind. Some dwarf ellipticals appear to form in the debris of major mergers and in particular in tidal tails, as do some globular clusters. Discs form and remain continuously revived in low-density regions by an ongoing supply of metal-poor gas. This comes from mergers of dwarf satellites, from which the gas is tidally stripped or ejected by dwarfs.

Supermassive black holes seem to be intimately connected to the formation of the central bulges or spheroids of galaxies. This is apparent from the correlation observed between the mass of the black hole and the spread of stellar velocities in the spheroid. The formation and feeding of these central black holes is likely to be driven by a collision or merger that drives gas into the galaxy core. One plausible idea is that the feedback on to the protogalaxy from an early quasar phase is responsible for the interconnection that self-regulates black hole mass. Black hole growth and the quasar phase should therefore be contemporaneous. Indeed, the number of quasars, and the number of merger-triggered ultraluminous infrared galaxies, rise rapidly in the past. The epoch of luminous quasar formation appears to coincide with the bulk of star formation in the universe. This peaked about eight billion years ago.

Black hole growth is certainly intimately connected to mergers, since mergers of gas-rich galaxies provide raw material for accretion on to and fuelling of black holes. The mergers of the central massive black holes provide the most plausible mechanism for understanding how the most massive black holes have formed. These weigh several billion solar masses or more.

Early in the history of every galaxy, stars were forming and dying at a great rate. Immense amounts of gaseous debris were generated. This gas fell into the centres of galaxies and fed the central black hole. This is how the black hole must have grown, by accumulation of matter from its surroundings. The firework display around the black hole would have been spectacular in our galactic youth. The black hole was undergoing a feeding frenzy.

This rather simplistic scheme for generating galaxy morphology and supermassive black hole growth via mergers has received support from both theory and observation. Simulations of major mergers between gas-rich discs demonstrate that tidal torques in the merging system effectively drive the gas into the central regions. The density is so high that the gas concentration is gravitationally unstable and is presumed to fragment efficiently into stars. Observations bear this out. Double radio jets are detected that require the presence of a binary system of supermassive black holes, a necessary precursor to merging. Radio-loud quasars thrive in the environments of rich clusters, where the supply of intracluster gas in the cluster cores may provide a reservoir of fuel for the supermassive black hole at the heart of the central giant galaxy.

The actual hydrodynamic simulation of a starburst is beyond current computational powers, but disc star formation can be modelled theoretically because of the observed correlation between the rate of star formation and gas density. Nature provides us with the conversion efficiency of gas into stars. Incorporation of this relation greatly reduces the computational requirements of a galaxy disc formation model. This correlation can be understood in terms of the gravitational instability of self-gravitating gas-rich stellar discs. Bulge formation is enhanced in galaxy clusters as a consequence of the tidal harassment suffered by gas-rich galaxies.

A more direct observational vindication of the merger model comes from its success in accounting for the observed counts and space densities of discs and ellipticals at various epochs in the universe. In fact, one cannot identify morphologies very easily, and one counts galaxies via their spectral energy distributions. Optical, infrared, and submillimetre counts can all be reproduced. We can also account for the star formation history of the universe and the diffuse background light from optical to sub-millimetre wavelengths. It would seem that mergers were a key element of our past history.

8 Clusters and Clustering

On the one hand, there are those youthful and enthusiastic but totally irresponsible cosmologists and theoretical physicists who build imaginary universes, which are neither of any scientific nor artistic value. These men simply lack the proper appreciation of the scarcity of definitely known facts and the realization that without such facts all speculation becomes largely futile. . . . On the other hand, there are far too many observers, especially some of those who have the use of the largest telescopes, whose knowledge of the fundamental physics is meagre. . . . Interpretations which are being made are all too often autistic rather than scientific in character.

Fritz Zwicky

To probe the past effectively, we need to study galaxies. These are fossils from the past, because their properties have changed little since formation. Clusters of galaxies are relatively pristine collections of galaxies. Clusters have not changed much since birth, and contain any debris that galaxies may have ejected in their youth. All galaxies in a cluster are more or less coeval, unlike a random collection of galaxies in the field. The study of galaxy clusters is an important step in our exploration of the past.

Galaxies aggregate together in clusters under the relentless pull of gravity. The Swiss astronomer Fritz Zwicky pioneered the cataloguing of galaxies and studies of clusters. He was an unconventional man, who occasionally had radical ideas, some of which have stood the test of time. It was Zwicky, for example, who proposed the existence of nuclear gremlins, the name he coined for what we now call neutron stars, by far the densest stars known to exist. A neutron star is a sphere of matter weighing as much as the sun but compressed into a region the size of Paris. The Sun spans half a million kilometres in radius and is at a mean density of water, about one gram per cubic centimetre. A neutron star contains this mass compressed into a few kilometres, and a teaspoonful of neutron star weighs about ten billion tonnes. Clearly, Zwicky had imagination. He was vindicated: our galaxy contains a billion neutron stars, distinguishable when young by their unique regular radio emission variations.

Zwicky also pioneered the notion of dark matter. He was responsible for demonstrating that clusters of galaxies predominantly consist of matter that astronomers cannot see. Clusters are the largest self-gravitating entities in the universe. They are important laboratories where cosmologists can test their speculations about what dark matter might or might not be. They are also immensely useful for mapping out the universe and its cosmic history, as we describe below. Zwicky rarely pursued his ideas to the point of quantifying them: dark matter and the clustering of galaxies were perhaps the exceptions. Many of his suggestions fell by the wayside, and Zwicky remained an outsider, beyond the pale of the astronomical establishment. He concluded one of his treatises on galaxy clustering by lamenting that 'after Pythagoras had discovered his famous theorem, the Greeks slaughtered 150 oxen and arranged for a feast. Ever since that happy time, however, whenever anybody proposed something drastically new, the oxen have bellowed.'

Understanding how gravity induces structure is the key to the clustering of the galaxies. Clusters are the largest self-contained systems in the universe. And they are youthful, indeed some are still in the process of formation. They are self-contained, supporting themselves with their own gravity, just as do galaxies. Unlike galaxies, clusters are relatively simple and unevolved systems. Clusters have retained much of the initial dark matter. Hence they are our best laboratories for probing cosmology.

Intergalactic gas in clusters is expected to be at a temperature of a hundred million degrees. It is measured in clusters via its X-ray emission. Galaxy clusters are found to contain immense amounts of hot diffuse gas – indeed, there is several times more mass in gas than in stars.

Galaxies also contain large amounts of interstellar gas, at least when they are not trapped in the gravity field of rich clusters. The interstellar medium may account for between 1 and 10 per cent of the mass in stars. It is enriched by gas ejected from evolving stars. In a cluster, most of the interstellar gas is swept out from the galaxies. Two effects lead to gas loss. One is the enormous pressure induced as the galaxies move at speeds greater than sound through the hot diffuse medium. Another is that supernova explosions of massive stars heat the interstellar gas so that much of it is ejected from the galaxy as hot wind. The ejected gas accumulates in the intergalactic medium. The gas within clusters is a dumping ground for the debris from its galaxies.

If the galaxy is near other galaxies, however, as is the case in a cluster of galaxies, then the gas reservoir in the halo is disrupted by interactions with the immediate environment. In a great cluster, frequent collisions occur between galaxies, but at relative speeds that are too large for mergers to occur. The stellar systems pass through each other like ghosts through a wall. But for the gas clouds, it is quite a different story. The clouds collide. The gas heats up and disperses into the intergalactic medium. Indeed, a great cluster of galaxies contains enormous amounts of hot intergalactic gas. The mass of gas is several times the mass in stars. The intracluster gas is gas that has been inhibited from forming stars. It stays hot for billions of years. We observe the diffuse gas as tenuous X-ray emitting plasma that pervades the intracluster space.

The gas in the clusters is at a temperature of a hundred million kelvin, emitting X-rays of several kiloelectronvolts energy. Despite this enormous temperature, comparable to that at the centre of the sun, the gas is at a pressure less than that of the most perfect vacuum attainable on earth. Yet the gas is spread over a region spanning millions of light years across. The mass of hot gas in a cluster amounts to a hundred trillion solar masses. To an X-ray observer, such clusters represent one of the most common sources in the sky to emanate from the distant universe. The gas is so hot because of the strong gravity field of the cluster. The galaxies, and the individual gas nuclei, have random motions of thousands of kilometres per second. For the gas, these random motions of nuclei simply manifest themselves as the gas temperature. We measure the motions of the massive particles, the galaxies, from the Doppler shift, and of the nuclei from the X-ray spectrum, which enables us to deduce the gas temperature. The random motions of the nuclei are in perfect agreement with the typical motions measured for the galaxies. Otherwise the gas, if it were too cold, would fall into the centre, or if too hot, would evaporate from the cluster. The gas is simply heated by gravity, the gas being compressed as the cluster formed. What is remarkable is the enormous quantity of gas. Galaxy formation is a very inefficient process. Only about 10 per cent of the gas was turned into stars.

Unexpected results emerged when the diffuse cluster gas was observed with the new generation of sophisticated modern X-ray telescopes. The hot gas was not a monolithic sphere of hot gas. Instead, the gas is highly structured in many regions. In the gas, one can see the evidence of past mergers between clusters. The mixing time of the gas is very long, especially in the outer regions. One often sees secondary structures that are off-centre. One sees clumps and filaments of hot gas that are relics of mergers. One sees giant bubbles, suggestive of heating by the radio lobes and associated shock fronts from a central radio galaxy.

Clusters are relatively young structures. Some are just forming at the present epoch of the universe. The crossing time across a cluster is billions of years or more. So with hindsight such structure is not surprising. The gas retains a memory of the past history. The gas temperature can vary from one side of the cluster to another, because there has not yet been enough time for mixing.

The diffuse cluster gas is found to be enriched with heavy elements that must have originated in supernova explosions. These generate and expel iron and other elements into the interstellar medium of the galaxies. As galaxies collide, the interstellar medium is heated and ejected into the cluster gas. Once the cluster gas accumulates, its pressure becomes much higher than that of the interstellar gas, which is only at a temperature of 10,000 kelvin or less. The cluster gas rams into the interstellar gas as the galaxy orbits the cluster. The ram pressure removes interstellar gas and further contributes to the intracluster medium. The net result is that the intracluster gas is enriched in heavy elements such as iron.

These elements are detected with X-ray telescopes. A nucleus of iron, which has only one electron in orbit, emits a line emission just like a hydrogen atom, except that the high charge of the iron nucleus means that the energy level of the electron is much higher than for hydrogen. X-ray telescopes detect the iron line emission, and we infer that the amount of iron in the intracluster medium amounts to about a third of what is found in the sun, relative to hydrogen. This is an enormous amount of iron by any standards. The gas has been thoroughly enriched, presumably by long-dead stars. Indeed, one requires three times as many supernovae to have exploded and dumped their debris into the intracluster medium than we would have expected simply by monitoring the abundances of the stars. Star formation and star death must have once occurred at far greater rates than observed today.

Weighing galaxy clusters

Even more dark matter than is seen in galaxy haloes pervades the intergalactic space in clusters. This dark matter can act as a gravitational lens, distorting the light from galaxies in the background. In this way, astronomers are able to chart the distribution of dark matter throughout the vast volume occupied by a cluster.

Clusters of galaxies are transparent. One can see distant galaxies in the background universe through the clusters. As the light from these galaxies passes through the clusters, it responds to the gravity field. According to Einstein's theory of general relativity, the space is curved by the mass of the cluster, so the path of the light rays from the background is curved.

If the background galaxy happens to be situated precisely behind the geometric centre of the cluster, its light is distorted into a ring, named after Einstein. The radius of the Einstein ring measures the strength of the cluster gravity field. In more realistic cases, the background galaxy image is arc-like. Several arcs are often produced. The arc-shaped galaxies are clearly not part of the cluster because they are found to have a redshift much higher, and hence a velocity of recession much larger, than that of the cluster. We are viewing a giant gravitational lens at work. Redshift is a measure of distance, and we infer that the lensed galaxies are far more distant than the cluster. By measuring the location and size of the arcs, as well as the redshift, we can infer the mass of the cluster.

Masses of clusters can be also derived by two quite different techniques. One just measures the redshifts of the cluster galaxies. Each is found to have a slightly different velocity along the line of sight to the cluster. The average is just the redshift of the cluster, but the spread of redshifts tells us about the random motions of the individual galaxies. These sample the cluster gravitational field. The galaxies are held together by the gravity of the whole cluster; otherwise the cluster would break up. We can use the motions of the individual galaxies, which tell us their kinetic energy, to work out what the mass of the cluster must be in order to hold them together.

Yet another approach utilizes X-ray emission from the hot gas that permeates the cluster. The X-rays are emitted by gas at temperatures of tens of millions of kelvin, and studying the X-rays allows us to deduce the the pressure of the gas. But the gas is in a state of equilibrium under gravity – it is neither collapsing nor expanding, and this balance again gives us a measure of the mass of the cluster.

All three ways of estimating cluster mass agree, and indicate that clusters are loaded with dark matter. Perhaps 2 per cent of the mass of a cluster is in galaxies and 10 per cent is in diffuse gas. The rest is dark and, since it is not detectable, cannot predominantly consist of baryons.

The hierarchy of clustering

We can see clustering even where it is not possible to recognize distinct systems. The painstaking procedure of counting millions of galaxies provided the breakthrough. Shane and Wirtanen at the Lick Observatory counted a million galaxies to nineteenth magnitude, out to a distance of half a billion light years. It took them ten years. Thirty years later, the counting process was automated with a laser-driven plate-measuring machine to allow two groups of astronomers in Edinburgh and Cambridge to count ten million galaxies in the APM (for automated plate-measuring machine) survey. The survey showed that the universe appears to contain a distribution of galaxies that has the same density everywhere and in every direction: it is uniform and isotropic. Is this an optical illusion?

From the spectrum of a typical distant galaxy, astronomers, most notably Vesto Slipher of the Lowell Observatory in Arizona during the early years of the twentieth century, deduced that distant galaxies seemed to be systematically receding from us. The spectrum was shifted towards longer or redder wavelengths relative to the spectrum of a standard laboratory source. As we will describe below, Edwin Hubble then found that the greater the distance to a galaxy the greater is its velocity of recession. This, as was realized later, was as a result of the expansion of the universe.

This universal law discovered by Hubble meant that to obtain the distance to a galaxy, we need only measure the spectrum of its light and determine the redshifts of any spectral lines. Redshift measures velocity of

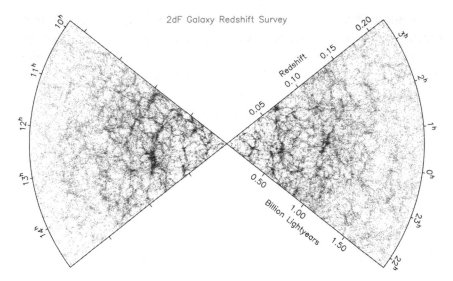

Figure 3 A galaxy redshift survey, completed with the two degree field camera of the Anglo-Australian Telescope.

Source: The 2dF Galaxy Redshift Survey Team.

recession – how fast the galaxy is moving away from us – which in turn translates into distance via Hubble's relation between distance and recession speed for the expansion of the universe. By following this procedure, three-dimensional surveys can be constructed. Current surveys utilize the measured distances to tens of thousands of galaxies. They confirm homogeneity on the largest scales, but also at smaller scales find voids and superclusters. To the limits of the largest telescopes (reaching to about twenty-ninth magnitude), astronomers estimate that there are a billion or more galaxies out to a depth of ten billion light years.

However, we need more than just images. We need spectra in order to obtain redshifts and take the measure of the universe. Indeed, we need millions of redshifts to perform definitive tests of the structure of the universe. As of 2006, one such survey has been completed of 250,000 galaxy redshifts, the Anglo-Australian Two Degree Field Survey, and a second, the Sloan Digital Sky Survey, is well on the way to its goal of a million galaxy spectra.

From gas to omega

A cluster is the largest self-gravitating object in the universe. It is a simple object, with only gravitational forces having played a role in its assembly. Most of the cluster mass, about 80 per cent, is dark. Of the rest, about 15 per cent of the cluster mass is in gas, and a few per cent is in stars. There is far more gas than stars within clusters.

We can assume that the amount of gas seen in a cluster, along with the stars, sums to give the total baryon content of the cluster and has been unchanged ever since the baryons were synthesized in the Big Bang. Now the actual density of baryons is known from our interpretation of the abundances of the light elements. We can exploit this situation, and deduce the density of non-baryonic matter in the universe.

It is convenient to express the average density of matter in the universe in terms of the critical value for a flat universe. We refer to the density, relative to the critical value for a flat universe, as omega (Ω). We infer that the density of non-baryonic matter in the universe is a third of Ω. If Ω were equal to one, the universe would be on the verge of collapsing in the future. If Ω were larger than one, the universe would collapse. If Ω were less than one, it would expand forever. In fact, it seems that Ω is about one-third. The universe will continue to expand forever.

Counting clusters

Such a dramatic conclusion needs to be verified. Counting galaxy clusters as we look back into the past provides such a test. A universe at critical density is finely balanced. In a given region, we expect the total energy to be conserved. When one throws a stone in the air, one finds that its total energy, determined when it is launched, does not change. Of course, its energy of motion is reduced and is zero at the maximum altitude of the trajectory. The gravitational potential energy increases to make up for the difference. It is the sum of kinetic and potential energy that does not change. For the universe, we find that the total energy is unchanged, and consists of two parts: the kinetic energy in galaxy motions and the gravitational potential energy. It turns out that these two energy contributions are nearly equal and opposite to each other. Indeed, in a flat universe, in which Ω is exactly equal to one, the total energy is exactly zero.

In such a universe, the kinetic and gravitational potential energies are in precise balance. This means that any local lump in density is unstable, as near the lump the gravitational potential energy dominates over the kinetic energy. The opposite would be true for a hole: it would be created by excess kinetic energy. The local excess in density accretes matter from its surroundings and grows more pronounced. A local deficit in density is drained of matter and becomes a void. This would not be the case in a universe that is well below the critical density, since the excess in kinetic energy over gravitational energy means that growth is retarded. However, even in such a case, a dense enough lump always overcomes the expansionist tendency of its surroundings, and accretion prevails.

This process of growth by accretion of mass from the surroundings is what we call gravitational instability. The instability is precisely what describes the growth of structures in the universe, and the increase with time in the number of structures that are larger than some given mass. Astronomers are finding galaxy clusters at great distances, corresponding to when the universe was half its present size. The numbers are small, but it would be virtually impossible to have any such distant massive clusters in a universe that is at critical density. In a universe at critical density, growth is too rapid, so there are too few massive clusters far away. The growth instability means that most of them are produced only recently: they would be unobservably rare at very great distance. But in a universe that is sub-critical the recent growth is retarded; hence the clusters must have formed earlier than in a universe at critical density. If Ω is about one-third, the cluster observations can be reproduced. Again, independent evidence points to a universe at sub-critical density.

Random motions of galaxies

The low Ω bandwagon continues to roll with the confirmation that another consequence of the development of clustering requires a low Ω universe. As galaxies cluster, they acquire random motions. If Ω were close to unity, the full effects of gravity make themselves felt, and the random motions are large. They amount to a thousand kilometres per second or more. Such large motions are easily measurable, since they bias distance determinations that are based on Hubble's law of expansion. The Two Degree Field Survey of galaxies with a quarter of a million measured redshifts has enabled astronomers to map the Hubble flow with unprecedented precision. The conclusion: there is a random component to the systematic expansion velocities of the galaxies. Astronomers traditionally refer to the random velocities as a peculiar motion relative to the systematic expansion of the universe. So-called peculiar motions are present but at a typical level of only about 300 kilometres per second. This can be understood if Ω is about one-third.

The intergalactic medium

We have seen that quasars are cosmic beacons that light up intergalactic space. They are the most luminous objects in the universe, conjectured to be a transient phase during the early stages of galaxy formation. By looking for traces of absorption in quasar light, cosmologists are able to detect intergalactic gas. A uniform intergalactic medium has not been discovered, but discrete clouds of gas have been detected in intergalactic space. Any diffuse intergalactic medium would cover all of space between the quasar and our galaxy. Near the distant, and highly redshifted, quasar, the absorption lines would also be strongly redshifted, but light absorbed by the medium closer and closer to our galaxy would be redshifted less and less. The result would be a continuous band of absorption. In contrast, discrete clouds of intergalactic atomic gas would produce narrow spectral lines.

Hydrogen atoms absorb light at a unique set of wavelengths. By measuring the characteristic wavelength, we can unambiguously identify hydrogen, and, for that matter, other chemical elements. Amazingly, the second most abundant element in the universe, helium, was first discovered in the Sun, by study of the solar spectrum. The most abundant cosmic element, hydrogen, generates the strongest such absorption feature or spectral line from the intergalactic gas that permeates the space between the galaxies.

This characteristic spectral line, known as Lyman alpha (α), is normally at a wavelength of 121.5 nanometres, in the far ultraviolet. This line would be impossible to observe with terrestrial telescopes in the nearby universe. If such radiation were to reach the surface of our planet unimpeded, it is unlikely that life as we recognize it would have survived. Fortunately for us, the Earth's atmosphere effectively screens out ultraviolet radiation. However, when we observe the quasars, the light is redshifted. The Lyman α line is shifted to longer wavelengths where it can readily be observed.

Such narrow lines at the Lyman α wavelength are indeed found in the spectra of distant quasars. They are produced in numerous discrete hydrogen gas clouds along the line of sight to those quasars, just as interstellar gas clouds produce absorption lines in nearby stars. Absorption lines from heavier elements are observed as well, but from the Lyman α line we learn that most of the absorbing material is in the form of the most abundant element in the universe, hydrogen. The absorbing gas is in discrete filaments, sheets, and clouds, and collectively adds up to an appreciable fraction of the baryons in the universe. The absorbing clouds show little contamination by heavy elements.

They are metal-deficient, and their metal fraction amounts to a hundredth of the abundance found for the diffuse gas in the great galaxy clusters. There is no evidence of any spatial clustering. This is suggestive of a relatively uniform distribution throughout space, one that does not favour the denser, more evolved regions of the universe. It would seem these clouds are likely to be the pristine material from which galaxies formed.

9 Space Is Nearly Flat

As lines (so loves) oblique may well
Themselves in every angle greet;
But ours so truly parallel,
Though infinite, can never meet.

Andrew Marvell

Our understanding of cosmology appears to be converging towards the idea that the universe is infinite. The geometry of the universe is close to Euclidean. Space is three-dimensional. In Euclidean space, parallel lines always stay parallel and continue to infinity. This means that in a two-dimensional analogy, the universe is as flat as a sheet. But a sheet, mathematically, is infinite. As goes the plane, so goes the universe.

In fact, we can never prove that space is precisely Euclidean or flat. The best we can hope for is to demonstrate that space approximates flatness. The universe may be very, very large, but we could never prove that it is infinite. Rather than try to measure its size, we can measure its geometry. This is a local measurement and one might imagine that it could be easier to implement, at least in principle, than a global measurement of space curvature.

Gravity is geometry

To understand the origin of this powerful conclusion let us go back to the nineteenth-century Russian mathematician Nicolai Lobachevsky. Lobachevsky pointed out that there are three possibilities for the geometry of a homogeneous and isotropic space. If space is flat (Euclidean), or negatively curved (hyperbolic), like the surface of a saddle, it must be infinite. Only if space is curved positively, like the surface of a sphere, can it be finite. Einstein's theory of general relativity predicts that gravity is in effect a curvature of space. Replace the gravity field of the sun by a slight curvature in the surrounding Euclidean space. The effect is that straight lines, as traced by light rays from distant stars, are no longer straight. Our precepts about geometry necessarily being Euclidean had to be abandoned. Parallel lines were no longer parallel. The deviations were small but measurable, and led to the prediction that gravity could now explain one of the most challenging problems in physics, the origin of the universe.

And as of the beginning of the millennium, incontrovertible astronomical evidence points to the near-flatness of space. The geometry of the universe is Euclidean. Space is flat. The result is written in the sky, where it has been measured in the form of a peak in the angular distribution of temperature fluctuations in the cosmic microwave background.

The Big Bang theory was a consequence of general relativity. The universe expanded in a precarious competition between the kinetic energy of the expansion and the gravitational potential energy that threatens to eventually initiate a contraction. If there is enough matter in the universe, gravity must dominate, the universe will decelerate and begin to contract. If the matter density is below a critical value, defined as Ω by cosmologists, the universe will expand forever. The critical density is well known and depends on the Hubble constant, currently measured to be within 10 per cent of 70 kilometres per second per megaparsec. But what is not known is the actual density of the matter in the universe, most of which is dark and hence exceedingly difficult to detect. General relativity predicts that if the universe is below the critical density, it will expand forever and the geometry of its space is negatively curved. It is said to be hyperbolic: in two dimensions, the geometry would resemble that of a saddle-shaped surface. A sub-critical universe that expands forever has positive energy. One that is above critical density has negative total energy, corresponding to the dominance of gravitational potential energy over kinetic energy. A critical density universe has zero energy. Einstein's theory links the energy of the universe with the nature of its geometry. Only a critical density universe is Euclidean.

Observations strongly hint that the matter density is sub-critical. Most, about 90 per cent, of this matter is dark, but its gravitational effects allow us to measure its density. We have seen that a variety of techniques point to a matter density that is one-third of the critical value, with an uncertainty of at most a factor of 2. However, we still cannot infer that the universe is destined to expand forever, because of a complication: dark energy, the source of a repulsive force that counteracts gravity. As Einstein showed, energy and mass are equivalent, and dark energy is in effect a source of negative mass.

Theory leads the way

Einstein originally introduced the concept of dark energy in the guise of the cosmological constant in 1917, to counter the gravitational effect of the matter and provide a static universe that neither collapsed nor expanded. As we have seen, the static universe was doomed to die. Hubble discovered that the universe was expanding in 1929. Theorists began the attack, although Einstein was initially not convinced. Neither Einstein nor, for that matter, Hubble could come to grips with the concept of an expanding universe. In 1927, on hearing from Lemaître about his new theory of an expanding universe, Einstein replied: 'Vos calculs sont corrects, mais votre physique est abominable' ('Your calculations are correct, but your physical insight is atrocious').[1] Nevertheless, Einstein was among the first to bow to data. In 1930, he became convinced of Hubble's expansion law, and he later admitted that the introduction of the cosmological constant term was one of the greatest mistakes of his life.

However, theory has a habit of rebounding. Pandora's box, once opened, is hard to close. Dark energy would never be forgotten, and the cosmological constant has occasionally resurfaced in order to account for some particular observational challenge that almost invariably faded as observations improved. The first real revival came from theory. In 1981 inflationary cosmology provided the first major new insight into the Big Bang since the 1920s. According to this theory, the universe underwent a profound change, a phase transition at 10^{-35} seconds, and expanded rapidly in scale for a brief period. This meant that the geometry of the universe was smoothed out and flattened. Inflation predicts that the universe is at the critical density. But, as we have seen, ordinary and dark matter could not account for the critical energy. The cosmological constant remains the plausible culprit, if indeed inflation occurred. But observations intervened.

The role of observations

If we could directly measure the geometry of the universe, we could by-pass the dark matter problem and test the inflationary prediction of flatness. Normally, scientific balloon flights last half a day. The balloon-born telescope is brought down before it has drifted out of range of the ground station. At the South Pole in particular, however, the circumpolar winds have allowed scientists to fly balloons for long duration flights of up to two weeks, eventually landing near the launch site. Long duration balloons have reached the sensitivity required to perform a satellite-type experiment, but at much lower costs.

Enter the BOOMERANG long duration balloon experiment. Designed to study the cosmic microwave background with unprecedented accuracy, this experiment was developed by an international group of astronomers led jointly by Paolo deBernardis of the University of Rome and Andrew Lange of the California Institute of Technology. Initially launched at the South Pole in 1999, BOOMERANG surveyed 2.5 per cent of the sky at an angular resolution of 15 arc-minutes, on a 10-day duration circumpolar flight. Theory predicts that there should be small temperature fluctuations in the cosmic microwave background, in order to seed galaxy formation. These were indeed discovered near the anticipated strength by the COBE satellite in 1992 at a much larger angular resolution of seven degrees. BOOMERANG was designed to search for the smaller scale signal that was crucial to our understanding of structure formation. The fluctuations were mapped for the first time.

More conventional balloons have also produced similar results. A comparable balloon package called MAXIMA, more sensitive but only flown in North America for a few hours, has obtained maps of the fluctuations at about 10 per cent the cost of the long duration balloon. However, the MAXIMA results appeared a week later, when the rival experimental results had already been acclaimed. No doubt the delay can be attributed in part to the low budget approach to science, by no means for the first time.

The much higher resolution of BOOMERANG compared to COBE has enabled a fundamental test to be made of the nature of the fluctuations. The primordial fluctuations are enhanced by the astrophysics of the early universe on small angular scales, of around a degree. This corresponds to how far a radiation pressure-driven fluctuation propagates in the early universe. This distance is limited by the age of the universe at last scattering, about 300,000 years. This so-called last scattering surface, or the horizon of the universe at last scattering of matter and radiation, has a physical scale of about 30 megaparsecs. The distance of the last scattering surface to us is about 6000 megaparsecs. From this, we infer that the characteristic angular scale is 45 arc-minutes in a flat universe. This enhancement, by about a factor of three, predicted by theory because of the effects of gravity, was measured by BOOMERANG. It constitutes a confirmation of the primordial origin of the fluctuations.

The fundamental result came, however, with the precise determination of the angular scale of the peak. The physical scale associated with the horizon of the universe at last scattering translates on the sky to an angular scale that depends on the curvature of the universe. If the universe is negatively curved, as in a lower density universe, the predicted peak shifts to small angular scales. In effect, the gravity field of the universe acts like a lens.

In fact the peak measured by BOOMERANG corresponds precisely to the expectation for a flat universe. The location of the peak indicates that the density is within a few per cent of the critical value. Nor was confirmation of flatness all that emerged from BOOMERANG. The predicted peak is at an angular scale of 45 arc-minutes, exactly the scale expected for the preferred flat model. BOOMERANG data continue, however, to 15 arc-minutes. Theory predicts a second high amplitude feature due to the wave-like oscillations of the radiation pressure-driven fluctuations, corresponding to the trough of the wave that crested at 45 arc-minutes. This feature has also shown up in the data as a smaller, second peak: it is smaller in amplitude than the first peak because the radiation redshifts vary slightly during the time it takes for the trough of the wave to be visible on the horizon scale of the universe.

The transient nature of astronomy was well demonstrated by the

BOOMERANG data. The first big surprise was that the second peak turned out to be lower than predicted. Within a week of the release of the first BOOMERANG data plots, the electronic Internet servers buzzed with speculations about why this might be so. The favourite among the preferred explanations is the idea that the baryon density might be as much as twice the value indicated by primordial synthesis of the light elements in the first few minutes of the Big Bang. Increasing the baryon density damps out the shorter wavelength sound waves, and reduces the amplitude of the peaks on smaller angular scales.

This is not a unique explanation, but it does lead to novel predictions. If the baryon density is doubled, as initial indications suggested, then the ratio of baryons to non-baryonic dark matter is also doubled. It ends up, in this case, being about 20 per cent. Now such a high baryon fraction means that baryon self-gravity plays a significant role. The oscillations, previously seen in the radiation, get transferred to the dominant dark matter by gravity. Galaxies are tracers of dark matter on very large scales. This leads to the possibility of a 'baryonic footprint' in the density fluctuations measured in galaxy surveys such as the 2DF and the Sloan Digital Sky Survey. In these surveys, baryon-induced oscillations are expected to become visible in the three-dimensional structure of the galaxy distribution on scales of the order of the horizon at last scattering, around 100 megaparsecs. Current data find significant evidence for such fluctuations. The effect appears to be present, albeit at a noisy level.

But again all changed with the next big surprise in the cosmic microwave background data. Here the charge was led by a relative newcomer to the field, the DASI experiment running at the South Pole, in spring 2001, led by scientists at the University of Chicago. DASI is a radio interferometer, comprised of 13 horns of 20 centimetres diameter. This experiment consists of an array of small telescopes that simulate a large antenna but have a resolution determined by the size of the individual horns. It synthesizes an aperture the size of the array, and measures the microwave background at the radio frequency of 30 GHz. To achieve the exquisite sensitivity of one part in 100,000 for temperature differences on the sky, a mountaintop site is required in order to reduce atmospheric backgrounds.

DASI reported fluctuations with a sensitivity greater than BOOMERANG, and found that the second peak was present at precisely the predicted strength. Very soon afterwards, BOOMERANG announced the revision of their earlier results, as more data were added, and everything fell into place once more.

In 2002, three ground-based interferometer experiments (DASI, CBI, and VSA) reported new high angular resolution measurements of the fluctuations in the cosmic microwave background. The angular distribution of the fluctuations agrees perfectly with the previous data and extends the data to smaller angular scales.

One experiment, the Very Small Array (VSA), is an aperture synthesis telescope consisting of 14 antennae each of 14 centimetres diameter sited at an altitude of 2400 metres in Tenerife and is led by scientists at the Cavendish Laboratory, Cambridge University. A second, the Cosmic Background Interferometer (CBI), led by a group of scientists from the California Institute of Technology, is an array of 13 antennae each of 90 centimetres diameter sited at 5080 metres on the Atacama plateau in Chile, one of the driest places in the world. The larger antennae mean that the CBI can achieve substantially higher angular resolution than the VSA. The conditions of high altitude and, on the Chile site, low residual atmospheric water vapour help to make the sites especially suitable for microwave astronomy by reducing the atmospheric background noise. Up to 100 square degrees of sky, well clear of the galactic plane, has been mapped in these experiments.

Cosmologists had already been basking in the glory of the BOOMERANG and MAXIMA microwave background experiments. The new experiments have higher angular resolution: while the balloon experiments probed angles of around 15 arc-minutes, the interferometers provide data down to about one arc-minute, giving us even more information on the detailed physics of the very early universe.

The previous experiments provided supporting evidence for the presence of these oscillations up to the third peak. The new experiments have confirmed this picture, with the CBI experiment extending the mapping to even smaller scales, showing the existence of a fourth oscillation at scales of a few arc-minutes, exactly as predicted by the theory.

The only surprise from CBI is the excess radiation near an angular scale of an arc-minute, which is slightly larger than the expected contribution from galaxy clusters. The intracluster gas up-scatters cosmic microwave background photons, producing a spectral distortion of characteristic shape. This results in a diminution of cosmic microwave background flux viewed through the cluster at frequencies below 150 GHz, and an enhancement at higher frequencies. This effect, named after Russian astrophysicists Rashid Sunyaev and the late Yaakov Zeldovich, has been seen for individual clusters, where it is relatively large, up to a thousandth of a kelvin. This is typically one part in 3000 of the cosmic microwave background. For the first time, however, CBI is reporting a detection of the integrated effect of all clusters in the line of sight, which at 30 GHz gives an effect of only 15 microkelvin. The effect is smaller than when one looks at a given cluster because the intracluster gas from the unresolved clusters fills only part of the projected beam in a random direction on the sky.

Enter the ACBAR telescope. Situated at the South Pole, this collaboration between Berkeley and Case Western Reserve University also measures fluctuations in the cosmic microwave background at arc-minute scales, but at high frequencies, between 150 and 274 GHz. It is an array of infrared radiation (or heat) sensitive detectors, called bolometers, designed to measure at submillimetre wavelengths on arc-minute angular scales. The additional frequencies are important, because the effect of hot gas scattering on the cosmic microwave background depends on the frequency at which one is searching. At 220 GHz, the effect vanishes, and at higher frequencies an actual enhancement is expected due to the energy input from the hot cluster gas. In effect, the Sunyaev–Zeldovich effect adds energy to colder photons, and converts them into hotter or higher frequency photons. The ACBAR data help to confirm the earlier results in detecting power from unresolved galaxy clusters.

A new discovery frontier was breached with the detection of polarization of the cosmic microwave background radiation. In 2002, DASI observed the first polarization signal. Polarization occurs because an electron scatters light in an intrinsically lop-sided way with a so-called quadrupole pattern. Even though the electrons are distributed randomly, they scatter light in all directions with random orientations of the electromagnetic wave oscillations that are the photons, but only if the light source is itself isotropic. However, in this case the source, the cosmic microwave background, has fluctuations, and in particular there is a quadrupole anisotropy. The net effect now is that electron scattering produces polarized light. The DASI detection means that one is probing the ionization history of the universe, since it is the free electrons that are responsible for most of the scattering. At late times the universe contained atomic gas, but at early times the gas was ionized. We are in effect probing the transition where most of the polarization is generated.

The next dramatic leap forward occurred in March 2003. We have described how a new cosmic microwave background satellite experiment, the first since COBE, obtained maps at unprecedented resolution of the entire sky. WMAP imaged the entire sky at radio wavelengths. COBE's resolution was seven degrees, WMAP's was a quarter of a degree. The precision of the new measurements is so high that one could now clearly view with high accuracy the three peaks in the temperature fluctuations on different angular scales, just as predicted by theory.

From the strengths and locations of the peaks and the damping tail, one could hope to extract ever more precise parameters of the cosmological model. The universe must be flat, as otherwise curvature of the light rays from last scattering would shift the peak locations. This fixes the total density of matter and energy in the universe. The universe is at the critical energy density that corresponds to spatial flatness, to within 5 per cent. The baryon density controls the ratio of odd to even peak strengths, since the baryons control the restoring force that governs the rarefactions relative to the crests. It is found to be 4 per cent, again to an uncertainty of 10 per cent, precisely the value inferred from primordial synthesis of helium, deuterium, and lithium in the first minutes of the Big Bang.

The matter density is determined from the absolute strength of the first peak. It is found to be 30 per cent of the critical density, to within an uncertainty of at most 20 per cent. The density fluctuation spectrum is controlled by the strength of the peak relative to the fluctuation strength on very large angular scales. The fluctuations in density are found to be just as inflation in its simplest guise predicted. And the damping tail of the fluctuations, the decline in peak strength to smaller angular scales, confirms a fundamental prediction of the theory. We are witnessing the trace of acoustic fluctuations imprinted on the cosmic microwave background, first predicted as long ago as 1967. It is these very fluctuations that persisted and grew in the dark matter, and eventually generated the large-scale structure of the universe of galaxies that we see today.

It seems to be inevitable in astronomy that each new discovery raises further challenges. The universe is flat, this we have learnt. Undoubtedly, this discovery will inspire new experimental efforts and new insights into the nature of the universe. The first of these revelations is with us. The universe is dominated by dark energy, the modern reincarnation of the cosmological constant.

Note

1. Cited in A. Berger (ed.) (1984) *The Big Bang and Georges Lemaître.* Dordrecht: Reidel.

10 Dark Energy and the Runaway Universe

I think that a vacuum is a hell of a lot better than some of the stuff that Nature replaces it with.

Tennessee Williams

It is contrary to reason to say that there is a vacuum or space in which there is absolutely nothing.

René Descartes

Supernovae are believed to be virtually perfect bombs. A supernova is brighter than a billion suns. When stars die, they brighten so much that individual stars can be detected even in distant galaxies. That suggests that supernovae might be good distance indicators. The best supernovae for this purpose are the brightest and most consistent. These are the supernovae of Type Ia, and they are generated by low-mass stars.

Exploding stars

On 24 February 1987, the first nearby supernova since 1604 was seen. It occurred some 170,000 light years away in the Large Magellanic Cloud. Its progenitor was a blue supergiant star 20 times more massive than the sun, and was itself highly luminous, attaining about 100,000 solar luminosities. It brightened a hundredfold during the explosion, and became easily visible to the naked eye. The heart of a massive star had imploded to form a neutron star, after exhausting its supply of nuclear fuel. As the matter compressed under immense pressure to form a ball of neutrons, immense amounts of energy were released, mostly in the form of neutrinos formed in the reaction: proton + electron → neutron + neutrino. These neutrinos were scattered by the inner layers of the progenitor star, and expelled in an immense explosion. Radioactive decay of unstable isotopes of cobalt, first to nickel, then to iron, continued to heat the expanding shell and power its optical luminosity for up to a year or more.

After the optical supernova, named 1987A, was seen, three groups of physicists received a wake-up call. They were manning deep underground facilities designed to act as supernova neutrino telescopes. After a re-examination of the data, neutrino detections were reported three hours before the optical explosion was noticed. Neutrinos interact so weakly that only a handful of events were recorded, but these sufficed to establish beyond any reasonable doubt that a neutron star had formed in the explosion. Many neutron stars are detected as pulsars, both radio and X-ray, and there are relics of ancient supernova explosions. The most recently formed pulsar is the neutron star seen in the Crab nebula, formed in an explosion in 1054.

Supernovae are subdivided into several types. The precursor stars that formed neutron stars were of between 10 and 20 solar masses. We refer to the resulting supernovae as Type 2, since they are the least luminous of the supernovae. The spectrum of a Type 2 supernova reflects the rich mix of elements that were ejected in the explosion. The precursor star's envelope would also contain material enriched in the core by thermonuclear burning, including carbon, silicon, oxygen, and iron, and these would be ejected along with large amounts of hydrogen and helium. All of this was ejected. There is also an intense neutron flux near the neutron star that irradiates the carbon and other elements to form much heavier species, such as uranium, barium, and europium. These rare elements are seen in old stars, and testify to their contamination by debris from supernovae in the distant past.

Stars below about 8 solar masses end up forming planetary nebulae, which expel most of their envelopes to form white dwarfs of about three-quarters of a solar mass. Such will be the fate of our Sun. Many stars are in binary pairs that eventually become pairs of white dwarfs. The merger of a pair of white dwarfs is a violent event. A white dwarf is made of carbon and oxygen, potentially excellent nuclear fuel except that, under usual conditions, the elements are under immense pressure in the compact white dwarf, with the atoms effectively in a crystalline lattice. Under these conditions, thermal energy is completely unimportant, and nuclear reactions cannot occur. But the situation is altered when a merger heats up the material. Carbon and oxygen can now burn via thermonuclear reactions with undiminished vigour to form iron. So much energy is released that the merged white dwarf is disrupted into expanding debris highly enriched in iron, with traces of carbon and oxygen. No hydrogen is present.

The stage is set for the most luminous variety of supernovae, Type Ia, produced by the collapse of a white dwarf star. A white dwarf becomes unstable to collapse if its mass exceeds 1.4 solar masses. This was a result first calculated by the Indian-American astrophysicist Subrahamanyan Chandrasekhar in the 1930s. His idea was so radical at the time that at first no one believed him. Such a collapse would be completely unstoppable. The star implodes under gravity, and the core initially forms into iron, the most stable of the elements in terms of its nuclear properties. But the collapse is so powerful that under immense pressure, the iron decomposes into neutrons, protons, and neutrinos. A large amount of energy is suddenly released, some of which is carried off by the neutrinos. The entire star explodes as a supernova. Theory suggests that the energy released by the collapse should not vary significantly for different Type I supernovae, since the amount of nuclear energy available is controlled by the maximum mass. The maximum mass a stellar core can burn before it collapses is about one solar mass.

These explosions are catastrophic, leaving nothing behind, other than a cloud of gas expanding at 7 per cent of the speed of light, and are characterized by the virtual absence of hydrogen in their spectra. Type 1a supernovae are up to ten times brighter than Type 2 supernovae, and their spectra are readily distinguishable by the dominance of iron and lack of hydrogen. They are one of the most common varieties of supernovae. In the process of disintegrating, some of the iron is converted into a radioactive isotope of nickel. The light is caused by energy released from the decay of nickel. This isotope is produced in a great and rather precise quantity during the collapse of the outer layer of the white dwarf core. The unstable nickel isotope decays to iron, and about seven-tenths of a solar mass of iron are ejected as the core explodes with a final burst of neutrinos. The spectra of the supernovae show that the ejecta consist predominantly of iron and other heavy elements. As a consequence, supernovae of this type should start off equally bright and then gradually dim at the same rate. The dimming is controlled by radioactive decay. We infer this because the predicted rate of decay with time of the luminosity of the supernova resembles that of radioactive decay. This means that the plots of decreasing luminosity with time for these supernovae are virtually indistinguishable in shape. The only difference is the apparent luminosity: the further away the supernova, the dimmer it is.

Yet a third type of explosion awaits stars of masses in excess of 25 solar masses. The core accretes so much mass during the final implosion that a black hole forms. Up to a hundred times more energy is released than in ordinary supernovae explosions. We call these events hypernovae, and they may be responsible for producing some of the very rare neutron-rich elements. Fortunately hypernovae are very rare, and observed to occur at no more than one-tenth the rate of supernovae, and possibly much less. One intriguing property of hypernovae is that they are considered to be associated with the most luminous objects in the universe, the gamma ray bursters. Intense bursts of gamma rays, requiring as much energy as is seen in a supernova, but lasting less than a minute, are associated with regions where massive stars form, and are seen at redshifts of 6 and even beyond. Potentially, the optical and infrared afterglows from these objects can be used to probe the host environments where the first massive stars are forming in the universe, at redshifts of 20 or further.

Type I supernovae have been observed in nearby galaxies of known distance, and it does indeed appear that not only is the rate of decay of visible light, or the light curve, identical, but the total luminosity emitted is also the same, to a good approximation. This means that we can work out the distances of far-off supernovae from their apparent luminosities.

Supernovae as distance probes

A supernova, especially when selected from its light curve and spectrum to be a certain type, always produces the same amount of energy. The luminosity peaks about one month after the star explodes. The peak optical luminosity is constant from supernova to supernova. Supernovae are so bright, attaining at peak the light of a billion suns, that one can see them at immense distances. Supernovae are therefore ideal tools for measuring the distance scale of the universe.

In practice, one needs to calibrate supernovae in nearby galaxies with other distance indicators so that one can pin down the local distance scale. Since supernovae in nearby galaxies are relatively few in number, one has perhaps half a dozen cases of nearby galaxies with a recent supernova. Only in these cases is it possible to measure some of the many variable stars, which provide very accurate distances because of their universal nature. The stars used for this purpose are the Cepheid variable stars, which can be studied both in the Milky Way and in the nearby galaxies. Cepheid variable stars have a distinctive correlation between their period of variability and their absolute luminosity, so that they form 'standard candles' whose distance can be measured accurately.

Once the half-dozen nearby galaxies with historical supernovae and monitored Cepheid stars have been calibrated, we can measure distances to galaxies that are at distances of up to a thousand or more megaparsecs, in other words a substantial fraction of the observable universe. Indeed, supernovae have been detected in galaxies beyond redshift unity, corresponding to a distance of about 10 billion megaparsecs. The supernova is typical of the explosion of a white dwarf star; that is to say, its spectrum is deficient in hydrogen. This is consistent with the properties of the host galaxy, which is an evolved, red elliptical. Such a galaxy today lacks enough recent star formation to provide the massive star progenitors of core collapse supernovae. The expansion derived by Hubble out to a recession velocity of 1000 kilometres per second has been extended more than 100 times further in distance scale using supernovae.

We can study supernovae so far away that it becomes practical to search for a deceleration of the distant galaxies relative to us. This, after all, is what is expected in the Friedmann–Lemaître model of the universe. If the universe were empty it would expand at light speed and always have the same speed as we looked further and further back in time. But add a modest amount of matter and the universe must decelerate.

However, a remarkable result emerged in 1998. Measurements of the distances to Type Ia supernovae at high redshift showed evidence for a dimming by about 20 per cent. After the more obvious suggestions such as dimming by dust have been eliminated, the simplest explanation is that this is most likely to be due to an acceleration in the expansion of the universe. If the expansion is accelerating, then a very distant galaxy with a measured redshift is at a greater distance from us than would be a galaxy of the same redshift in a uniformly expanding empty universe. It would be at an even greater distance from us compared to the more realistic case of a decelerating universe. To see the effect, one has to look back in time to when the universe was half its present size. And even there the difference in distance is 10 per cent, which produces a 20 per cent change in the light received by us.

If there is a cosmological constant, then the result is a kind of anti-gravity, and there is acceleration. The observed acceleration is predicted for a universe that is spatially flat but in which two-thirds of the critical density is accounted for by the dark (or vacuum) energy associated with the cosmological constant. For such a universe, the age is approximately $1/H_0$ (H_0 is Hubble's constant), or 15 billion years.

A modern view of the cosmological constant associates it with the energy density of the vacuum of space. At a quantum level, the vacuum is not empty at all but a seething mass of virtual particles that appear and disappear in pairs in times too short to be measured, as predicted by the uncertainty principle.

As the virtual particles and antiparticles arise and disappear in pairs, there is no change in the charge. But there must be an effect associated with the density of energy, since the quantum motions are a form of pressure and energy. The bizarre and non-intuitive result is that the pressure of the vacuum is negative.

There must be such an energy associated with a vacuum that is teeming with virtual particles. Squeeze a gas of virtual particles and the pressure decreases. There are fewer virtual particles to contribute to the pressure since the volume of the region determines how many there are. Make the volume small enough and the finite uncertainty in distance that Heisenberg requires means that there are none left. The vacuum energy has been measured in the laboratory. The ground state energy of atoms is very slightly perturbed by the presence of the vacuum. The energy of the vacuum is found to have negative pressure.

Einstein's theory of gravity requires energy and pressure to act as a source of gravity. Normally, pressure in a collapsing cloud of gas at first opposes collapse. Near the final state of a black hole, pressure actually helps the collapse. Ordinary pressure is a source of attraction in extreme gravitational fields. But with a vacuum the opposite is true. Vacuum energy has negative pressure, and so acts repulsively. Vacuum energy is like antigravity. This is why acceleration is a unique prediction of the cosmological constant.

The detection of acceleration via the dimming of distant supernovae is an important result. If it is to be accepted, all other explanations must be carefully excluded. For example, could the dimming of high redshift relative to low redshift supernovae be an artefact of the intervening dust? Normal interstellar or atmospheric molecules and dust preferentially extinguish and scatter blue light relative to red light. For this reason, the sun at sunset is red, and the sky is blue. However, no colour differences are found, and one would have to appeal to a weird and unique form of extinction, in which dust indiscriminately took out light at all wavelengths. Even uniformly absorbing dust makes little sense. The further one looked, the more variation would be expected in supernova peak magnitudes, since the dust would not plausibly be distributed uniformly in space. In fact, astronomers observe no apparent variation in supernova peak magnitudes as they study more and more distant galaxies.

A more serious concern is the possibility of evolution of the supernova itself, so that they may be untrustworthy as distance probes after all. Indeed, theory suggests that there may be wide variations in supernova properties, in which case the high redshift observations that sample a large volume of the universe might well be vulnerable to bias. Yet local supernovae are observed in a wide range of environments, where there are predominantly old as well as young stellar populations, and as we have noted earlier, no systematic differences in absolute luminosity are found.

The supernova results require dominance of dark energy over matter, and are in concordance with the cosmic microwave background inference of a flat universe at critical density. They also correspond with the observed cluster abundance, which fixes the matter density at about a third of the critical value. Cosmologists are happy, since a consistent cosmological model beckons. With independent verification of the key unexpected parameter, dark energy, from two totally independent experiments, one has every reason to believe we are approaching the final solution to cosmology. Only one of these, the supernova experiment, actually directly measures acceleration, the main clue to the vacuum energy. The microwave background tells us that something else other than 'cold' dark matter is needed, and this is most likely to be dark energy.

Dark energy

Cosmologists have now gone full circle, ending up with a value of the cosmological constant about 30 per cent smaller than Einstein originally introduced for the static universe. One can interpret the cosmological constant as a constant energy density of the vacuum that has only recently begun to dominate the mass density of the universe. It isn't possible to observe any such energy directly, which is why it is referred to as dark energy. The matter density decreases as the universe expands. When the universe was about one-quarter of its present size, the dark energy first became comparable to the matter density. One consequence is that the universe switched from deceleration under the influence of the gravitational attraction of matter to acceleration under the influence of the gravitational repulsion of the dark energy.

Dark energy is bizarre stuff. It is uniform, and always stays uniform. It does not cluster like ordinary matter under the influence of gravity. All that dark energy possesses is energy density and pressure. As we have seen, negative pressure is repulsive and causes the universe to accelerate once it is dominant. Dark energy, then, accounts for two-thirds of the mass–energy density of the universe.

There is no complicated explanation for dark energy: it can simply be regarded as a contribution to the energy of the vacuum. Dark energy is completely uniform, and remains so. It is only detectable via its effect on the acceleration of the expansion of the universe. Dark matter, however, *does* cluster. And this suffices to keep the astronomers very busy indeed.

11 The Panacea of Cold Dark Matter

O dark dark dark. They all go into the dark,
The vacant interstellar spaces, the vacant into the vacant.

T. S. Eliot

There is but one right, and the possibilities of wrong are infinite.

Thomas Henry Huxley

Most of the matter in the universe is dark. In contrast to dark energy, we can 'see' dark matter because it clumps as structure forms. Dark matter is detectable. And it amounts to about a third of the total mass–energy density of the universe.

The dark matter budget

The cosmic mass budget is best expressed with respect to the critical density for a universe that is spatially flat, the Einstein–de Sitter model. This density only depends on Hubble's constant, which is now known to reasonable precision. The critical density is 200 billion solar masses per cubic megaparsec. It is useful to compare this number with the density of starlight. We can express this in units of the luminosity of the sun, and it amounts to a hundred million solar luminosities per cubic megaparsec. If the universe were at the critical density, the ratio of mass to luminosity would be 1000 solar masses per solar luminosity. This gives a clear prediction for the closure of the universe.

What is actually measured is far less. Galaxy clusters gave the first indication of the prevalence of dark matter on large scales as early as 1933.

The first reliable values, however, came from galaxy rotation curves, which show the rate of rotation as we go further and further out from the galaxy. This provided proof of dark matter dominance in ordinary galaxies, and in particular in our own Milky Way galaxy. The rotation curves for large spiral galaxies are generally flat at large distances, indicating that the mass surprisingly increases with increasing distance from the galactic centre.

Galaxy rotation curves are measured via radio techniques using characteristic features in the spectrum, in particular the 21 centimetre line of atomic hydrogen, and in the optical band by other emission lines of hydrogen atoms. The radio method is very sensitive to low density atomic gas but lacks resolution. On the other hand, the optical measurements have excellent resolution but only track dense gas clouds ionized by massive stars. The combination works well. Consistent results are obtained, and dark matter is found to be ubiquitous on scales of up to a hundred kiloparsecs. In galaxy clusters, great progress has been made since the early determinations that used the kinetic energy of the cluster galaxies and compared it to the potential energy. The cluster mass was estimated from the mass needed to prevent the galaxies from flying apart and the cluster dissolving. The random motions of the cluster galaxies are inferred from the optically measured spread of their radial velocities.

We have previously described the two independent techniques that confirm the dynamical measurements of cluster masses. One utilizes X-ray measurements of the hot intracluster gas. The gas is at a temperature of around ten million kelvin. Its temperature is measured via X-ray spectroscopy. The gas has two opposing and balanced forces on it – the dispersive force of gas pressure balancing gravitational force associated with the self-gravity of the cluster. Assuming that the gas is in this 'hydrostatic equilibrium', we can then infer the cluster mass.

The other approach makes use of gravitational lensing by clusters of remote background galaxies. Gravitational lensing distorts galaxy images. An exactly aligned spherical lens, which could be an intervening galaxy cluster or a massive galaxy, converts the background galaxy image into a ring. More typical alignment and lens geometries result in several concentric arcs rather than a ring. The separation of the arcs is a measure of the dark matter content of the lens. All three methods measure the distribution of dark matter on large scales, and consistently yield a value of 300 solar masses per solar luminosity. The scale probed is several millions of light years. On larger scales, we run out of structures that are held together by their own gravity. There are no gravitationally bound structures in equilibrium that can reliably be probed. One method uses galaxies that are still receding from the Virgo Cluster of galaxies, but whose recession is being slowed by gravity. These galaxies are eventually destined to fall into the Virgo Cluster. Measurement of the infall motion of galaxies in the Virgo Supercluster region, the greater Virgo Cluster, probes the dark matter density on scales of up to 20 megaparsecs.

Another probe of the dark matter density on even larger scales, up to 100 megaparsecs, makes use of the fluctuations in the counts of galaxies obtained in large-scale galaxy redshift surveys. The clustering of the galaxy distribution on large scales is measured by fluctuations in the galaxy counts averaged over randomly placed spheres. The matter on large enough scales must be correlated with the light. The fluctuations inferred in the matter density provide a gravitational source that induces perturbations in the Hubble expansion. These are observable as random motions of galaxies and of galaxy clusters. Similarly, one can study the influence of agglomerations of dark matter on the Hubble expansion. Were the universe at critical density, large Hubble flow distortions, galaxy-peculiar velocities, and cluster streaming motions would be observed, amounting to a thousand or more kilometres per second. But the observed random motions of galaxies amount to about 300 kilometres per second.

The observed variations in Hubble flow indicate a value of the mass-to-light ratio that is equivalent to 300 solar masses per solar luminosity, or about a third of the critical value for a closed universe. This is in approximate agreement with the mass-to-light ratio inferred for galaxy clusters. A similar inference is made from examining the very small image distortions produced in the images of high redshift background galaxies. These can only be seen statistically, and the phenomenon is known as weak gravitational lensing.

Yet another method utilizes the redshift evolution of the number density of clusters. The rich cluster abundance above a given mass is observed to increase only slowly as the universe expands. The theory of cluster formation predicts a rapid increase of the massive cluster abundance in a critical density universe, due to the growth of density fluctuations driven by gravitational instability. This effect is systematically suppressed in the recent past if the density of the universe is below the critical value.

The bottom-up universe

The *ab initio* approach to large-scale structure has met with great success. One starts with infinitesimal density fluctuations in cold dark matter on very small scales. Inflation boosts these to scales that correspond to the horizon scale at the epoch when densities of matter and radiation were equal. This epoch coincidentally amounts to that of the scale containing the mass of a cluster of galaxies. Once the universe is matter-dominated, the fluctuations are gravitationally unstable and accrete surrounding matter. This results in galaxy cluster formation just before the present epoch, with galaxies forming when the universe was about a tenth of its present size.

Formation of structure occurs in a bottom-up sequence. Matter fluctuations are larger on smaller scales. This gives gravity a boost. Small objects condense first, and then larger and larger objects condense. This need not have been the case. One can equally well imagine a situation in which formation was top-down. It all depends on the initial density fluctuations from which structure grew. In fact observations strongly favour a bottom-up view of the universe. Galaxy clusters, the most massive self-gravitating objects, are at much lower density and are dynamically younger than galaxies. The mean density reflects that of the universe when they formed. Galaxies certainly formed before clusters. This is the essence of bottom-up theory.

Bottom-up structure formation has been extensively simulated, usually in the context of a cold dark matter dominated universe. The theory is well formulated for dark matter, and has been modelled via numerical simulations. Small clouds represented by aggregations of point masses gather under the attraction of gravity and merge into ever larger assemblages. A hierarchy of structure develops, as not all substructures are erased.

The successes of hierarchical structure formation are numerous. It can account for galaxy clustering. Simulations of large-scale structure are indistinguishable from actual surveys. On the largest scales where effective comparison is made, up to hundreds of megaparsecs, one can measure the distribution of density fluctuations in the galaxy distribution. The structure and its texture are naturally generated in a flat, dark energy dominated universe over these scales. The abundance of galaxy clusters is an effective probe of large-scale structure. The luminous matter traces the dark matter. The random motions of the galaxy clusters, relative to the universal expansion, trace inhomogeneities in the dark matter. The light that we see in the galaxies, if converted into dark matter with a universal constant of proportionality, exactly accounts for the random motions of the clusters as well as the observed accelerations of galaxies in the clusters. There is no need for progressively more dark matter relative to luminous matter on the largest scales. There is a need for dark energy, but this is only discernable on the horizon scale, gigaparsecs away. Dark matter is an unbiased tracer on large scales and requires a universe with a matter density equal to about a third of the critical value. And these top-down hierarchical models indicate galaxy formation at redshifts of up to 10, when the universe was a tenth of its present size.

To simulate is to approximate the cosmological truth

The patterns of structure formation are sufficiently complex that one has to resort to numerical simulations in order to confront many of the observations. The success of the theory is demonstrated by its accounting for several properties of the observed universe. Particular highlights are the strength and distribution of the cosmic microwave background fluctuations and the properties of galaxy clusters. Simulations of galaxy cluster formation succeed in reproducing the shapes and density structures of galaxy clusters.

In a typical simulation, the collapse of a massive cloud of gas is followed under the action of its own gravity. Initially, the cloud is supported by gas pressure, but the gas loses energy as it radiates freely. The cloud continues to collapse and forms a concentrated ball of gas that is glowing in X-rays, since its temperature is tens of millions of kelvin. The cloud is embedded in a cluster of thousands of galaxies and accounts for about 10 per cent of the cluster mass.

Such numerical simulations can reproduce the texture of the observed X-ray emitting intracluster gas. On larger scales, intergalactic gas permeates

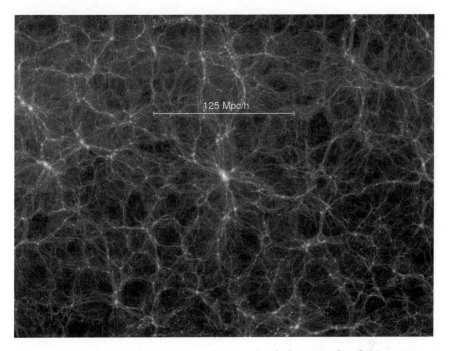

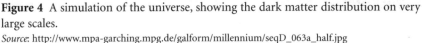

Figure 4 A simulation of the universe, showing the dark matter distribution on very large scales.
Source: http://www.mpa-garching.mpg.de/galform/millennium/seqD_063a_half.jpg

the universe. We can observe this intergalactic medium because it absorbs light from distant quasars. It is distributed in ionized hydrogen filaments and lesser inhomogeneities, and in vast, galaxy-mass clouds of atomic hydrogen. By including gas in the simulations, we have obtained a convincing explanation of the distribution of intergalactic gas.

However, galaxy formation has proven to be less tractable. Cold dark matter lumps merge hierarchically to form dark haloes of galaxies. Some substructures survive, but high numerical resolution is needed to follow the interactions and evolution of the clumps. Star formation requires implementation of yet another level of complexity, in order to allow for feedback of energy into the gas from forming and dying stars. The computational challenges are more than current computers can adequately handle. For progress, compromises must be made, and then one is never quite sure of the reliability of the results.

Putting on a spin

Why do galaxies rotate? This was one of the first questions raised when astronomers first resolved the beautiful spiral nebulae some two centuries ago. Rotation seemed an obvious property to explain the patterns. However, it required modern dynamical measurements to confirm this via sophisticated optical and radio frequency mapping of the gas in spirals.

Optical light comes from associations of massive young stars, whose ultraviolet radiation ionizes the surrounding hydrogen cloud in which the stars were born. The resulting emission, when analysed with a spectrometer, is found to consist predominantly of spectral lines from glowing hydrogen and oxygen gases. The expected wavelengths of the lines are known from laboratory measurements for gas at rest. The observed spectral lines are displaced to both the red and the blue, depending on which side of the galaxy one is observing, relative to the centre of the galaxy. This is due to the Doppler shifts resulting from the rotation of the galaxy: gas moving away from us on one side of the galaxy is shifted to the red, while gas on the other, moving towards us, is shifted to the blue. A similar technique is applied to radio emission from interstellar clouds of atomic and molecular gas, and yields similar results.

The spin of the galaxies was produced when they were still gas clouds contracting to form the first stars. The clouds have very irregular shapes. As a consequence, neighbouring clouds exert a pull on one another that causes each cloud to tumble or rotate, in such a way that no net spin or angular momentum is actually created. These forces are tidal torques in action between neighbouring clouds that generate an initial angular momentum for each galaxy.

It was at first thought in the 1980s that during the subsequent contraction of the forming galaxy, the angular momentum would be conserved. Just as a pirouetting dancer spins faster when bringing her arms in close to her body, the contracting cloud would spin up. The contraction stopped when the centripetal force balanced gravity: a rotating disc had formed. In this way, one would account for the sizes of the observed spiral galaxies, which are basically differentially rotating discs of stars.

With the advent of complex numerical simulations with high resolution in the late 1990s, it has become apparent that most of the angular momentum lost by the condensing gas is effectively transferred outwards into the halo. Dense baryon clumps sink into the central regions of the forming galaxy and lose angular momentum. Clumpiness in the dark halo also tidally perturbs the disc and aggravates the angular momentum transfer problem. The result of the simulations is that the disc sizes are invariably too small by a factor of about five.

The simulations have revealed another problem. The distribution of dark matter in low surface brightness dwarf galaxies, which are everywhere dark matter dominated, cannot readily be explained. This is studied via the rotation of the galaxies, and generally soft cores are found. The dark matter simulations invariably find a strong central concentration, which looks quite unlike the inferred dark matter content of the observed galaxies. The theory fails to explain the observed cores of the low surface brightness dwarfs. It may be that resolution of this problem just requires more detailed, higher resolution simulations that include the full interactions between the baryonic matters and the dark matter. So far, this has been lacking.

Not all the news is bad. Simulations (including gas) of merging galaxies show that the transient stages of mergers can look like very irregular galaxies. We see the development of tidal tails and spiral arms. While it is not generally possible to predict the nature of such objects in any fundamental way, one can set up initial conditions appropriate to the onset of a merger and compare the outcome, followed on a computer, with observed examples. Nearby examples of merging galaxies are relatively rare. In the early universe, we expect that mergers were far more frequent.

Mergers are an inevitable outcome as the hierarchy of structure develops. We learn a lot about human mortality by studying sick people. Similarly, the pathological cases of irregular and disturbed galaxies are often the most interesting. As mergers would have been more common in the remote past, we can search for them by inspecting very deep images taken with a telescope in space, where one has the resolution to detect the tidal features predicted in ongoing mergers. Another approach is to search for the energy release that must inevitably be triggered by the merger. A merger produces a luminous radio galaxy or a strong burst of star formation.

Luminous ellipticals have more mass in stars and have a stronger gravitational field, on average, than do luminous spirals. Their haloes are consequently more massive. One might expect them to have formed recently. Yet ellipticals have predominantly old stellar populations. This difficulty is resolved by assuming that major mergers, the precursors of ellipticals, formed stars efficiently over the time it took the merging galaxies to form a single system. This happens relatively rapidly for a close encounter in little more than an orbital crossing time. The gas is compressed and cools, loses its kinetic energy, and falls into the inner regions of the merged system. Here it efficiently forms stars and little gas remains.

This results in the formation of a spheroid of stars, as in the bulge of a spiral or an elliptical galaxy. In contrast, the discs of spiral galaxies form by a far more gentle process. Discs are cold and fragile. Slow infall of small gas clouds, or the occasional minor merger with a dwarf galaxy, provides the reservoir of gas that forms the disc. Gradual accumulation of gas into discs, in low-density environments, results in a continued gas-rich star formation at low efficiency for a Hubble time. Such a scheme may be said to 'work', in the sense that ellipticals are red and spirals are blue.

Astronomers use the largest telescopes to peer back in time and study the youthful phases of galaxies. One can in effect see evolution in action. The morphologies of galaxies change in the distant universe, especially as revealed with the exquisite resolution of the Hubble Space Telescope. The high redshift galaxies fit well into a hierarchical formation scheme. Field studies show a significantly increasing population of blue, irregular galaxies towards fainter magnitudes. These galaxies dominate the very faint galaxy counts. All of this is consistent with early formation via mergers and strong tidal interactions. Observations of the distant galaxies may be considered to represent a qualitative success of the hierarchical galaxy formation theory.

Most baryons are dark

Only about 10 per cent of the dark matter in the universe is baryonic. Modern determinations of the abundances of helium, deuterium, and lithium are found to be consistent with a Big Bang origin. Now that the cosmic microwave background blackbody temperature has been measured accurately, the light element abundances provide an accurate accounting of the primordial baryon abundance.

There are three further independent confirmations of the baryon fraction in the universe. Cosmic microwave background radiation fluctuations measured the baryon density when the radiation was last scattered by the matter, some 300,000 years after the Big Bang or at redshift 1000. Other determinations come from studies of the intergalactic medium at two distinct epochs. At high redshift, when the universe was only a quarter of its present size, one sees clouds and filament of intergalactic neutral hydrogen in absorption in the spectra of quasars. One needs to apply the ionizing photon flux, measured directly via the quasar emission spectra, to infer the total amount of intergalactic gas. At low redshift, we measure the hot intracluster gas in galaxy clusters via its X-ray emission flux to be about 10 per cent of the total cluster mass. Since clusters are considered to be sufficiently massive to have preserved their original baryon content, one can also deduce the baryon content of the nearby universe. All methods agree, to within the uncertainties. We find that the baryon density is about 4 per cent of the critical density.

There are problems, however. We can only account for about half of the predicted baryon fraction today in known sources such as stars and diffuse intergalactic gas. As many baryons as we see glowing in emission, whether in stars or in hot or cold gas, are missing from the accounting of the baryon budget performed today. About half the baryons that once were present are simply not seen directly. In addition to the dark matter problem, which refers to non-baryonic matter, there is also a dark baryonic matter problem. And, finally, there is the issue of the supermassive black holes. Lurking in the cores of galaxies, they contribute little to the dark matter budget but may profoundly influence galaxy formation.

12 Origins

Evolution . . . is a change from an indefinite, incoherent homogeneity, to a definite coherent heterogeneity.

Herbert Spencer

Some call it evolution, and others call it God.

William Carruth

To astronomers of antiquity, and indeed until the first three decades of the
twentieth century, the universe was static. It was inconceivable to imagine
it being in a state of expansion. Even Edwin Hubble, who discovered the
law that relates distance to redshift, refused throughout his life to accept
that the universe was expanding. But Hubble notwithstanding, there was a
paradigm shift. Too many facts were explained by the hypothesis of the
expanding universe. It became mainstream cosmology within decades
after Hubble's law was announced. How did this radical change in our
view of the cosmos come about?

The Big Bang unveiled

The idea that, in the mean, the universe should be uniform and isotropic,
enshrined as the cosmological principle, led to a remarkable simplification
of the equations that described the gravitational field. In 1917, Einstein
found a static cosmological model that could only be prevented from
collapsing under the relentless tug of gravity by invoking a repulsion force.
This force was enshrined as the cosmological constant, a term that has no
counterpart and no effect in Newtonian gravity, but is important only on
cosmological scales. In modern parlance, we identify the cosmological
constant with the energy of the vacuum. This is the so-called dark energy,

uniform everywhere in space, and responsible for the observed acceleration of the universe.

In 1917, however, there was no reason to believe in anything but a static universe. In fact, Einstein had overlooked the only true cosmological solution to the field equations that satisfied the cosmological principle and that did not require the introduction of a cosmological constant. His mistake was soon rectified.

The story of the Big Bang begins with a Russian mathematician by the name of Alexander Friedmann. He worked as a meteorologist, and his observations were made from high altitude balloons. At one point, he held the world altitude record for manned flight. Friedmann died at the age of 36 in 1925, supposedly from pneumonia that developed after a data-gathering flight. But two years earlier, in 1923, Friedmann had discovered something Einstein had overlooked: the possibility of the expanding universe. Working independently, a young Belgian priest, Georges Lemaître, came to similar results in 1927.

Within three years, Edwin Hubble, a former advocate who was bored with law, succeeded in measuring the distances to nearby galaxies. Velocities had been measured earlier by Vesto Slipher at the Lowell Observatory in Flagstaff. He observed the spectral lines characteristic of galaxy spectra and due to light absorbed by atoms in the atmospheres of the many stars that cumulatively produce the spectrum of a galaxy. He noticed that these lines were systematically shifted to redder wavelengths in many nearby galaxies, compared to the wavelength of atoms at rest. This spectral shift is a manifestation of the Doppler effect, previously encountered in our discussion of the measurements of stellar motions. It also yields the recession velocity of an entire galaxy.

Hubble monitored variable stars during the 1920s at the Mount Wilson Observatory. He succeeded in using their average brightnesses as distance standards to infer distances to galaxies. Distance can be measured if we identify a class of stars that all have identical luminosity. In practice, no stars are like this, but variable stars are the next best thing, if we can relate the variability timescale to their luminosity. A class of luminous variable stars, Cepheids, with a distinct pattern of variability provide ideal distance measurement tools. These luminous giant stars swell and contract in a series of periodic pulsations. The luminosity increases as the star swells, and the mean luminosity is found to be proportional to the period of its oscillation.

The astronomer Henrietta Leavitt undertook a painstaking survey of Cepheid variable stars that were all located in the Large Magellanic Cloud. She was able to establish the relation between luminosity and period, since these stars were all equidistant from the Sun. Many Cepheids were identified throughout the Milky Way. This relation was then used to establish the size of our Milky Way galaxy and, subsequently, the distances to our nearest neighbour galaxies.

Hubble's astounding result was that the further away a galaxy, the faster it receded from us. A galaxy, however, is a self-gravitating object and is perfectly stable. As space expands, the centre of mass of the galaxy moves, but the galaxy is so overdense locally that its interior remains unaffected. A galaxy is sufficiently dense relative to its surroundings for its own self-gravity to overcome the tendency of space to expand locally. Einstein's theory led to the interpretation that space is expanding; galaxies themselves are not!

The expanding universe was much later dubbed the Big Bang by the cosmologist Fred Hoyle, arch proponent of the singularity-free steady state theory, for the simple reason that it expanded from a point-like singularity of infinite density. At the outset, it was realized that this singularity was a mathematical artefact indicative of missing physics that was only supplied half a century later. In this way, according to Einstein's theory of gravity, space itself was uniform, unbounded, and expanding. There was no centre and no edge to space. The implication of the expansion of space is quite incredible: everything once started out from a singular state of extreme density.

The birth of physical cosmology

The Big Bang theory predicted the expansion of the universe, a result that in the early decades of the twentieth century was sufficiently radical that many, including Hubble himself, never fully accepted it. Galaxies have been rushing away for 14 billion years in a fiery explosion. What did cause the paradigm shift to an expanding universe? Confirmation came in several stages. There are now three pieces of evidence for the Big Bang theory that attested to an origin remote in time and emanating from an incredibly dense and hot state.

Of these, the first and foremost was the expansion of the universe. No other explanation carried any real weight. The universe is expanding, hence it was once in a highly compressed state. In the 1970s, Stephen Hawking and Roger Penrose showed that a past singularity, or state of near infinite density, was inevitable, given our then understanding of gravity. Only the possible existence of a cosmological constant provided a loophole in their reasoning.

The evidence for the expansion of space gradually became more and more convincing. In hindsight, it is remarkable that a young postdoctoral researcher who was a virtual outsider in the ranks of professional astronomers, Georges Lemaître, formulated one of the greatest predictions of modern physics, that the universe should be expanding, into a relation that expressed the proportionality between the recession velocity of a distant galaxy and its distance. In 1929, Hubble verified the redshift–distance relation. This soon became enshrined as Hubble's law of proportionality between recession velocity and luminosity distance.

Hubble used the brightest stars in more distant galaxies as his basic distance indicators. He explored a region that extends to the Virgo cluster of galaxies. With hindsight, we know that Hubble's distance indicators were partly erroneous, since he could not distinguish regions of ionized gas from stars. We also know that random motions dominate the region between us and the Virgo cluster, where Hubble's galaxies were located. The uniformity of the universe only becomes manifest beyond Virgo. Nevertheless, Hubble in 1929 announced his discovery of the redshift–distance law. The redshift was produced by the Doppler effect and resulted in a systematic displacement towards longer wavelengths for a receding galaxy. Blueshifts would be indicative of approach; only a few of the nearest galaxies have blueshifted spectra.

As we have seen, the prevalence of galaxy redshifts had been discovered in the first decades of the twentieth century. The fainter the galaxy, on the average, the larger its redshift. However, the observers who tried to understand the relation between distance and redshifts paid too much attention to the theoretical cosmologists, who only knew about the possibility of redshifts in the de Sitter universe. The de Sitter cosmological model was a strange beast. It was an empty universe in which the distance increased rapidly with redshift. The nearby galaxies in this model displayed a dependence of distance on redshift such that the distance to a galaxy increased proportionately to a non-linear function of its redshift. To his credit, Hubble did not care a great deal about theory. For him, the data reigned supreme. He re-evaluated distances more precisely than his predecessors had done, and inferred the linear relationship that we know as Hubble's law. To his dying day, Hubble remained reluctant to accept that the universe was expanding, despite the predictions of Lemaître and of Friedmann before him. However, within years of Hubble's announcement, most of the cosmological community seized upon Hubble's law to infer that space was expanding.

It is difficult in retrospect to understand how Hubble inferred a linear law, given the enormous uncertainties in galaxy distances and the fact that Hubble only initially sampled such a small volume of the universe. Hubble's constant (H_0) is measured in units of velocity per unit distance, in effect an inverse time. Hubble inferred a value of 600 kilometres per second per megaparsec. The modern value of H_0 is smaller by an order of magnitude, amounting to 70 kilometres per second per megaparsec, with an uncertainty of about 10 per cent. The inferred timescale $1/H_0$ is a measure of the age of the universe if no deceleration (or acceleration) has occurred. The age inferred from Hubble's measurement was around 1.5 billion years, and far less than the known age of the Earth. Hence many astronomers were at first reluctant to accept the expanding universe interpretation.

What changed? First, the cosmologists were very ingenious. Under the influence of Lemaître, the cosmological constant, first introduced by Einstein to make the universe static, was reintroduced. A hybrid model received much attention. Eddington and Lemaître advocated a universe that began from a static phase that would last as long as necessary before beginning to expand. Lemaître showed that galaxies could form in such a universe. A variant was an expanding universe that underwent an extended coasting phase as a consequence of the effect of the cosmological constant, with expansion eventually taking over. Such approaches psychologically aided the transition from static to expanding universe. And of course they greatly extended the age of the universe.

Most significantly, however, the observers revised the distance scale. This came about in part by recognition of Hubble's significant error in confusing the brightest stars with giant clouds of ionized gas, so-called HII regions. Allan Sandage from 1960 onwards was primarily responsible for developing a new distance calibrator that made use of the brightest stars and HII regions in galaxies as standard candles. This enabled him to probe the universe out to great distances and to reduce Hubble's constant to below 100 kilometres per second per megaparsec.

A major breakthrough came in the 1950s when Walter Baade recognized that there are two types of Cepheid variable stars. The confusion between the distinct types was only removed when Baade succeeded in identifying two distinct populations of stars, each with associated Cepheid variables, in the Andromeda galaxy. He realized that there were two types of Cepheids, which differed appreciably in luminosity. Baade was able to double the distance scale. The remaining improvements came more slowly. For nearly 40 years, cosmologists debated Hubble's constant within the range 50–100 kilometres per second per megaparsec.

Final resolution came when the Hubble Space Telescope was able to resolve Cepheids in several galaxies outside our Local Group, in which supernovae were also found. As we have seen, Type 1a supernovae are uniform in their properties, and can be used as standard candles, thus allowing the distance scale to be expanded dramatically. One could now use the same measuring 'tool' to connect the distances within our galaxy and its neighbours to galaxies as remote as 15 megaparsecs. The previous measurements of the Hubble constant were refined, and the result was the modern value of 70 kilometres per second per megaparsec.

Another major advance was the realization that the helium abundance is universal. The fraction of helium is essentially the same wherever we look and in an amount that far exceeds the quantity ordinary stars could have synthesized. The helium must have had a pregalactic origin, and the early Big Bang is the logical environment. The helium could only have been produced at the very high densities obtained in the Big Bang if the universe was once very hot, hotter than the centre of the Sun, so that thermonuclear reactions converted hydrogen to helium. In fact, some deuterium and a trace amount of lithium would also have been produced in the first three minutes. This seemed highly plausible, but perhaps not inevitable, until the third link in the cosmic jigsaw puzzle fell into place. It was George Gamow, the Russian-born physicist, who asserted that the universe began hot and that it was at this time that the chemical elements were synthesized by nuclear reactions. He proved only partially correct – only the lightest elements were produced in the Big Bang, as we have just seen – but his work led collaborators Ralph Alpher and Robert Herman to predict that the present temperature of the universe should be about 5 kelvin. A decade elapsed before the discovery of the relic radiation field by Arno Penzias and Robert Wilson in 1964, which attested to the existence of a primordial cosmic furnace. The relic radiation is universally known as the cosmic microwave background radiation, because most of the energy in this radiation is at microwave frequencies.

The primordial fireball

The fossil radiation was expected to have the spectral energy distribution of a perfect blackbody – an object that absorbs and emits all wavelengths. This would reflect its origin in the ideal furnace of the Big Bang. In 1990, the perfect blackbody spectrum was confirmed by the Cosmic Background Explorer, or COBE for short, satellite: a temperature of 2.726 kelvin was measured. Precisely the same temperature was measured in all directions. No deviations from this temperature could be detected to less than a hundredth of a per cent. The blackbody radiation must have been produced in a cosmic furnace: the universe once was hotter than hell!

The relic radiation is the best blackbody known to man. Its blackbody spectrum is indisputable proof of an origin in a dense and fiery past. Blackbody radiation is the ultimate level in loss of information content. All is random, all order has gone. Take, for example, a Rolls-Royce automobile alongside a Rent-a-Wreck. Pulverize the cars, and one still has distinguishable material content between the two. But put them into a sufficiently hot furnace until all that remains are protons, neutrons, electrons, and radiation. The radiation has attained a blackbody state and communicates energy freely with the particles and its environment. There is no remnant of the information that went into the furnace. Blackbody radiation is the ultimate in democracy: all sources are equivalent and indistinguishable, apart from the temperature.

Today, radiation travels freely to us from distant galaxies. Conditions in

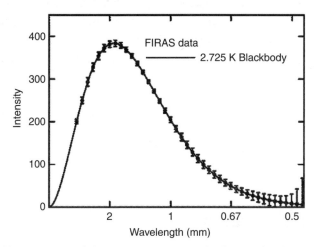

Figure 5 The spectrum of the cosmic microwave background radiation. The error bars represent 400 standard deviations.

Source: J. Mather *et al.* (1994) *Astrophysical Journal,* vol. 420, p. 439.

intergalactic space are far removed from producing blackbody radiation. Such radiation requires intimate contact between matter and the photons. This occurs, for example, in an ideal furnace or at the centre of the Sun. In the very early universe, too, any radiation could travel barely any distance at all before being immediately absorbed and re-emitted. The matter and radiation were in thermal contact. Their temperatures were identical. A blackbody is at a single temperature. Complete thermal equilibration is precisely what is needed to create blackbody radiation. The very early universe must have resembled the interior of a furnace.

By tracing the progress of the expansion backwards through time, we can predict the mean density of the matter in the universe at any epoch, and in particular long before galaxies had formed. From the measured blackbody temperature, we can now extrapolate to any density in the past. We can deduce what the universe must have been like at different epochs in the past. Only in the first few months after the Big Bang was the universe in the form of a plasma that was dense and hot enough to have created blackbody radiation. Although radiation continues to interact with matter by scattering off of free electrons, photons are not absorbed and new photons are no longer created. However, the radiation produced in the first year remains as blackbody radiation: no process can destroy it. The expansion of the universe results in the gradual cooling of the radiation temperature, but the spectrum retains the form of a blackbody. The cosmic background radiation is a direct relic of the primordial fireball, generated within a matter of months after the Big Bang, and has no other remotely plausible explanation.

The age of the universe

Type Ia supernovae have been detected out to a look-back time of half of the present age of the universe, from when light was redshifted by a factor of 2 in wavelength. The distance measurements are precise enough (to 15 per cent) that acceleration of the universe has now been confirmed. A perfectly linear law only applies if there is no acceleration or deceleration. Deviations from Hubble's linear law are found for the most distant supernovae. The measured age of the universe, inferred from Hubble's constant and the measured acceleration, is 14 billion years.

There are two completely independent measures of the age of the universe. Radioactive dating via thorium and uranium isotope measurements is applied to the abundances in old halo stars. Both thorium-232, with a half-life of 14 billion years, and uranium-238, with a half-life of 4.5 billion years, have been detected in two halo stars, measured with the world's largest telescopes. Nuclear astrophysics theory provides an estimate of the initial abundances relative to iron. The observed ratios provide an estimate of the time elapsed since the explosion of the supernova that synthesized these elements. This is already a major fraction of the age of the universe. We need only add the time taken for the galaxy to have formed and the supernova to have exploded since the Big Bang in order to deduce the age of the universe. The explosion ejected the heavy elements mixed into the supernova debris. This was eventually incorporated into the molecular clouds from which stars like the Sun formed.

Another age determination comes from application of the theory of stellar evolution to globular clusters. Globular star clusters are systems of millions of stars that predate most of the stars in our galaxy. We know they are old because the abundances of metals as measured in stellar spectra are low compared to those in the Sun. Hence the globular clusters must have formed long before the Sun.

As stars radiate energy by thermonuclear burning of hydrogen into helium, they evolve in luminosity. They become brighter as the fossil fuel is gradually exhausted and the central temperature rises. Heavier elements are burnt, first helium, then carbon, to provide the central temperature and pressure. Once the nuclear fuel supply of hydrogen, helium and carbon is exhausted, the star soon runs out of fuel.

If the star initially weighed less than 8 solar masses, its final fate as the core heats up is that its envelope swells. The star becomes a luminous supergiant. The outer shell is expelled to become visible as a planetary nebula. The ejecta slowly, after some 10^4 years, fade away, and a white dwarf is all that remains. If the star initially weighs more than 8 solar masses, its central pressure builds up to a level that the star core implodes via neutron capture and neutrino emission. A neutron star forms in the core and the release of binding energy drives a supernova explosion of the massive star precursor variety, or Type II.

In a globular cluster, the stars formed coevally. Here one has a snapshot of stars of different masses, which have reached differing evolutionary points. One can thereby infer the age of the globular cluster from comparison with models of stellar evolution, our best estimate being 13 billion years. To this must be added about one billion years for the delay between the Big Bang and the formation of the globular cluster, to give an age for the universe of some 14 billion years. Remarkably, this agrees well with the independently determined ages from the expansion of the universe and from uranium/thorium decays.

Our thermal history

The discovery of the microwave background radiation led to some remarkable insights into the beginning of the universe. The Big Bang was once a fireball. Only after the universe expanded to about one-thirty-thousandth of its present size did it become matter-dominated. Why this epoch? The expansion factor is just the ratio at present of the matter to the radiation density. This ratio specifies the corresponding redshift, which is identical to the expansion factor and applies to the stretching of the wavelength of light. In the past, radiation dominated the density of the universe, since the radiation density falls off more rapidly with time than that of the matter. The fall-off is by just one factor of (one plus) redshift due to the fading in energy (or increase in wavelength) of individual photons.

Only in a universe dominated by matter could the density fluctuations be gravitationally unstable and grow in strength. For the cosmic microwave background to have attained the intimate thermal contact with matter needed in order to generate its blackbody spectrum, it must have had a dense past. The radiation became blackbody, and remained blackbody, as long as there were sufficiently frequent interactions of the photons via scattering off of free electrons. More dramatically, under classical general relativity, the universe must inevitably have undergone a past singularity. However, all bets are off on such theoretical inferences about a past singularity if there is a cosmological constant. More precisely, a cosmological constant is equivalent to a cosmic repulsive force that, while far too small today to be of interest at the beginning of the universe, might nevertheless have once been far stronger.

One can now begin to reconstruct the thermal history of the universe. Quantum gravity is how it all began. Quantum gravity supplants general relativity within 10^{-43} seconds of the Big Bang. Physicists refer to this moment as the *Planck instant*. It represents the point in time before which the general theory of relativity breaks down, and quantum effects must have been dominant. The temperature was then about a tenth of a billion trillion trillion (10^{32}) kelvin. To simplify matters, we express temperature in terms of proton rest mass energy, one unit of which is a billion electronvolts (1 GeV for short). A trillion (10^{12}) is a thousand billion, and 1 eV is equivalent to a temperature of 10,000 kelvin. The Planck temperature now becomes a mere ten million trillion (10^{19}) GeV. There is as yet no preferred theory for this regime, which ordinarily is considered to be well beyond the reach of any feasible particle accelerator machine. However, higher dimensional theories of quantum gravity include models in which Planck scale physics is manifest at tera (trillion) electronvolt energy scales. As the universe expanded and cooled below the Planck scale, the ensuing evolution can be sketched out as follows.

The physics story, which we can probe, really begins at a temperature of about a hundred thousand trillion trillion (10^{29}) kelvin. The universe was then at an energy of ten thousand trillion gigaelectronvolts. The universe passed through this moment rather rapidly. Indeed this temperature was only maintained for 10^{-35} seconds. Nevertheless, before this time the energy was so high that the fundamental forces, which are totally disparate at low energy, were all identical and indistinguishable. They had merged together. The electromagnetic and weak and strong nuclear forces were all of equal strength. This was a time of great symmetry, when neutrinos were indistinguishable from neutrons, or whatever their predecessors were. This moment is called the era of grand unification.

In fact, there are four fundamental forces, all very distinct from one another at the present epoch in the universe. There is the electromagnetic force, the two nuclear forces, weak and strong respectively, and the gravitational force, which is by far the weakest. Today, the electromagnetic force, which holds electrons in atomic orbits, is a hundred times weaker than even the weak nuclear force, which holds neutrons in nuclei. But as the temperature of the universe increased beyond the point at which nuclei broke up into their constituent quarks, the electromagnetic force grew stronger and the nuclear forces grew weaker. Soon, the electromagnetic force rivalled the weak nuclear force, at 10^{-12} seconds. At much higher energy, only 10^{-35} seconds after the Big Bang, the nuclear force fell into line: all atomic and nuclear forces were indistinguishable. These are predictions of the quantum theory. A mist of quantum fluctuations shields the electron's intrinsic strength yet amplifies the nuclear forces that emanate from quarks.

As the temperature dropped below 10^{16} GeV, the new forces came into play that had no pre-existing counterpart. The strong nuclear force now dominated over the other forces, and the symmetry of grand unification was spontaneously broken. The resulting change in quality of the matter content of the universe involved the transient appearance of a new type of energy field. This field, dubbed the inflaton field, was responsible for the subsequent inflation of the universe. The inflaton is analogous to latent heat in a phase transition such as the melting of ice, when energy is released. Fish survive in frozen lakes for this very reason.

Einstein's equations show that if the density of inflaton field energy dominates (this is an invisible form of energy, much later baptized as dark energy), it remains constant as the density of ordinary matter drops because of the continuing expansion. Because the energy density is constant, the universe begins to expand at an exponential rate. Indeed, the universe continued to expand exponentially as long as the inflaton field was the dominant source of energy density. This phase of inflation began when the universe was about 10^{-36} seconds old.

This energy eventually decays away (by design) and inflation ends by about 10^{-35} seconds. The enormous kinetic energy turns into heat, and we are now again in the conventional hot Big Bang phase, initially dominated by radiation and relativistic particles. The inflaton is similar in some respects to the cosmological constant, except that its energy density was larger by about 120 factors of 10.

Now let us fast-forward our cosmic movie strip to a much lower temperature. At 100 GeV, the universe is a ten-billionth of a second old. Below this energy, the electroweak forces decouple from the strong nuclear forces that hold nuclei together. The fundamental force strengths subsequently resemble those we see today, with the nuclear forces being strong and short-range compared to the feebler electromagnetic force and the vastly weaker gravitational interaction. At this epoch, there is another less dramatic change in the state of the matter, as the different nuclear forces decouple. This change in phase of the universe helps to generate a small asymmetry in the baryon number, the number of particles minus antiparticles.

The baryon number measures the net amount of matter in the universe relative to antimatter. It is expressed in dimensionless form as the difference between the number of particles and antiparticles, divided by the sum. The sum corresponds to the number of photons in the microwave background, since at high enough energy, photons create particle–antiparticle pairs. As the temperature drops further, all the strongly interacting particles annihilate into radiation. Over time, the radiation redshifts to become the cosmic microwave background. There are one billion microwave background photons for every proton in the universe. Very few pairs of particles and antiparticles survive, because of the strong interactions that annihilate almost all of the pairs. Any relic particles froze out of thermal equilibrium once the temperature dropped sufficiently. The relevant temperature corresponds to the energy at which a fraction of the particle mass converts into pure energy (via Einstein's equation $E = mc^2$).

There are virtually no relic proton–antiproton pairs that survive. Today the baryon number is only 10^{-9}. But this number shows that clearly some baryons survived without antibaryon counterparts. To account for this value, there certainly must be a net baryon number, which guarantees that some baryons survive. These baryons became the present matter content of the universe. The antimatter content is less than a hundredth of a per cent; otherwise one would see gamma rays resulting from matter–antimatter annihilations. The observed universe consists almost exclusively of matter.

So far we have focused on heavy particles like protons. More generally, we call them hadrons. The hadrons consist of baryons and mesons. The fundamental components of baryons are quarks: each baryon is composed of three quarks, and each meson consists of two quarks. But matter is also composed in part of light particles or leptons. The electron is a fundamental particle that is a lepton.

Just as with the hadrons, there is also a net number of light particles today, the leptons, as estimated relative to the number of antileptons. Leptons include the electrons and their antileptons, the positrons, as well as muons. The difference between baryon and lepton number is always conserved in the late phases of the universe. There are no processes that mix baryon and lepton number in the low energy universe.

The baryon number itself becomes diluted with time. But so does the lepton number. It turns out that the number of leptons per baryon is related to the number of photons per baryon. This is because at early epochs and high temperatures, leptons, most notably the electrons and positrons, annihilated to produce the radiation content of the universe. Now a measure of the entropy per particle is the number of photons per baryon.

Photons are also generated at a later time, when galaxies form. Early on, annihilation into photons also increases the entropy. We infer that entropy increases, as is inevitable according to the second law of thermodynamics. However, it is very difficult to change the number of baryons relative to the number of leptons. For particle physics, the baryon–lepton number difference is generally sacrosanct. The challenge is to explain why it was non-zero. If it were zero, almost all of the baryons would have annihilated. The solar system would not have formed. Our presence requires a non-zero baryon number or a non-zero lepton number. One guarantees the other.

There is a particle physics approach to understanding the problem of why the observed universe is so matter-dominated. This involves tinkering with baryon number conservation. Conservation laws such as this are ordinarily sacrosanct, and can only be varied at two possible instants. One is at the epoch of grand unification of the electromagnetic and nuclear forces. This is when baryon number is ordinarily laid down. But another moment of vulnerability in terms of lepton number conservation occurs much later, at the moment when the weak force separates from the electromagnetic force. This occurs at an energy of 100 GeV, 10^{-10} seconds after the beginning. Since lepton and baryon number are inextricably linked, the asymmetry of the baryons, and the accompanying leptons, can be imprinted at either epoch. The jury is out on this one.

Dark matter particles are another relic from nanoseconds after the Big Bang. We have described how elusive they are. Heavy weakly interacting neutral particles are a favoured candidate for the dark matter. If such stable weakly interacting particles are indeed present in the universe, they would have frozen out in substantial numbers regardless of whether there was any primordial asymmetry. The lightest supersymmetric particle or neutralino is such a possible stable relic.

A priori, we have no idea about its interaction strength. However, a clever idea gives us the answer. If it interacts too strongly, too few survive to the present. Essentially all are annihilated at very high temperature. If its interactions are too weak, too many survive. One would have too much dark matter. The neutralino turns out to be a viable candidate for the non-baryonic dark matter in the universe only if its interaction strength is typical of that of the weak interaction. It is a weakly interacting massive particle, and expected to weigh in at a hundred to a thousand proton masses. Armed with these insights, we can now appeal to theory to justify our expenditures on the various indirect and direct detection experiments that we previously described.

Nuclei would fly apart were it not for particles called gluons that keep protons together in the confined space of the nucleus. Gluons are to quarks what muons are to protons and electrons. They help to bind quarks together to form protons, just as muons are responsible for binding electrons to protons to form neutrons. All the chemical elements are made of neutrons and protons. The universe was once a soup of quarks, gluons, electron–positron pairs, neutrinos, and photons. At a temperature of about 200 million electronvolts, yet another phase transition occurred when the quarks and gluons formed hadrons. The universe now contained protons and neutrons in thermal equilibrium, and with an abundance of approximately one neutron for every ten protons. The predicted number of neutrons depends only on the mass difference between proton and neutron. Once the temperature drops below a million electronvolts, the neutron-producing reactions stop, and neutrons freeze out.

At half a million electronvolts, the pairs of electrons and positrons annihilate and neutrinos freeze out. The neutrinos remain today as a hitherto unobservable background sea. The neutrons, however, have a much more important destiny. Neutrons are gobbled up by hydrogen to form helium and other light elements. At these high temperatures, the stage is set for nucleosynthesis of the light elements. This commences at a hundred thousand electronvolts or a temperature of a billion kelvin, when deuterium nuclei can first form by combination of neutrons and protons. Subsequent reactions produce He^3, He^4, deuterium, and Li^7, all of which are generated in abundances that are measurable today in primordial environments. The lack of stable nuclei at masses 5 and 8 means that nucleosynthesis peters out after He^4 is synthesized. The predicted primordial helium-4 incorporates nearly all the neutrons. The result is that there is about one helium nucleus for every ten hydrogen nuclei. By mass, this means that helium accounts for about 25 per cent of the mass of the universe.

Deuterium and the baryon fraction

The abundances of deuterium and helium are used to measure the total abundance of baryons in the universe. These include all the baryons that are not today found in stars. We do not directly measure the total abundance of baryons, since most of them could be dark. However, we can infer their abundance from indirect considerations of light element production long ago. If the baryon abundance at the present epoch were higher, the early nuclear reactions, which occurred when the universe was hot, would have been more efficient. More helium would have been synthesized, at the expense of deuterium, less of which survives. The relic helium and deuterium abundances probe the total abundance of baryons present in the first three minutes.

One expects to find primordial helium in such unprocessed environments as the intergalactic medium, the outermost parts of galaxies, metal-poor galaxies, and even, after some chemical processing and fractionation, in meteorites and the atmosphere of Jupiter. All abundances are consistent with a unique density of baryons, amounting to 4 per cent of the critical density. This is one of the best known numbers in cosmology. The uncertainty in this estimate is only 10 per cent. The universe remained dense and hot enough for thermal equilibrium of matter and radiation to be maintained until an epoch of about one month. This was when the cosmic blackbody radiation was effectively generated. One can study this epoch by searching for slight deviations from the blackbody spectrum of the cosmic microwave background. Any spectral distortions would probe the physics of the universe at this epoch.

The temperature continued to drop. The hydrogen was ionized and the radiation scattered frequently. The billion photons for every baryon sufficed to keep the hydrogen fully ionized until the temperature dropped below a couple of thousand kelvin or about 0.2 eV. At this point there were too few photons with energy above the hydrogen ionization threshold of 13.6 eV to keep the hydrogen fully ionized. The protons and electrons combined to form hydrogen atoms. Unlike free electrons, hydrogen atoms are very poor scatterers of electromagnetic radiation. Scattering of the photons stopped abruptly. The universe became transparent to the cosmic microwave background radiation. This occurred 300,000 years after the Big Bang.

Matter and antimatter

The laws of physics are symmetric with regard to matter and antimatter. Antimatter particles have the opposite charge to particles of matter. An antiproton has a negative charge, while a proton has a positive charge. The atomic mass and spin are identical, however. Antihydrogen atoms consist of a positron orbiting an antiproton. If a distant galaxy were made of antimatter, it would not look any different from a normal galaxy. The atomic energy levels are unchanged, the spectrum is unchanged. An atom of antihydrogen has identical spectral lines to an atom of hydrogen. Spectral lines are produced by antiatoms in the atmospheres of antistars, and one could not distinguish spectroscopically between stars and antistars.

But there would be profound observational implications if an antigalaxy were to exist. Ordinary matter would surround it. Intergalactic space is permeated by diffuse gas. Atoms of matter would come into contact with the diffuse interstellar medium and stars of antimatter. The result would be catastrophic: atoms of matter and antimatter annihilate upon contact. The annihilation of a proton with an antiproton produces gamma rays. These are highly penetrating and can be seen even if emitted at the far frontier of the universe.

Gamma rays are actually blocked by the Earth's atmosphere, but not by interstellar clouds or by intergalactic clouds. Satellite experiments have been flown with gamma ray telescopes on board. Any antigalaxy lurking within our horizon would produce so much gamma radiation in its environment that one would surely already have seen it. No such gamma-ray source has been found. There are clearly no antigalaxies and no antistars within the observable universe. This is one less hazard for intergalactic travellers of the future.

The five phases of creation

Let us review our history of the universe. We cover a vast span of time, 14 billion years of cosmic expansion. There are five major episodes in cosmology when the rate of expansion of the universe undergoes a change. These are the moment of the singularity, the epoch of inflation, the period of radiation domination, the domain of matter domination when large structures form, and the epoch of the late acceleration when the cosmological constant dominates the energy density. The cosmological constant describes the physics of antigravity, a mysterious and exceedingly weak force that was originally introduced by Einstein to stop the universe from collapsing. The original motivation has long since vanished, now that we know the universe is expanding, but we are confronted with new observations, to be described in a later chapter, that are compelling us to revive Einstein's original idea of a weak repulsive force.

General relativity, Einstein's theory of gravity, has an even more serious omission. It does not incorporate quantum theory. Gravity theory, as we currently understand it, must therefore break down at the beginning of the universe. New physics, quantum gravity, is needed. The quantum phase is at the beginning, at instants after the initial singularity. The cosmic clock starts ticking at the singularity, time zero, but the first 10^{-43} seconds is the realm of the new physics. The universe expands in a regime where quantum physics is omnipotent, and our physics is not reliable or even applicable.

The door is open for speculations about how physics is capable of dealing with the most extreme conditions imaginable, as the initial singularity is approached. Deep inside but ordinarily inaccessible to any observer is a singularity in space and time. Our best hope so far for handling singularities has come from Einstein's theory of general relativity. The theory predicts the existence of black holes. The cosmologist Stephen Hawking conjectured the existence of very small black holes, which evaporate into a cloud of energetic particles and photons. Some experts view the approach to the singularity as being heralded by a quantum foam of such mini black holes appearing and disappearing as they self-destruct by a process of evaporation.

Nor is this all. There is nothing special about the number of space dimensions. Perhaps we were once in a higher dimensional space. It is tempting to think that just as Einstein curved three-dimensional space to incorporate gravity, one might need to invoke other curved spaces in higher dimensions to account for the properties of the elementary particles. This led to the idea that space might initially have been four- or ten-dimensional. At the end of the quantum era, the extra dimensions may have curled up. Our universe is one point of space–time in some higher dimensional super-space. We can only ever detect such a higher dimensional superspace via gravity, if at all.

Higher dimensions are hard to visualize, and hard to justify until we have a theory of quantum gravity. The next phase of the expansion is easier to grasp because it involves classical physics. The universe spontaneously underwent a phase transition. This would have happened when certain types of particles disappeared. As the temperature dropped,

(a)

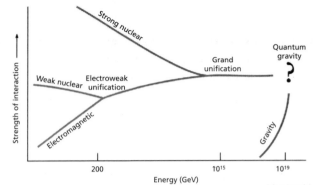

(b)

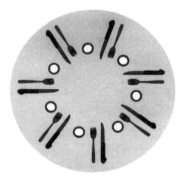

Is your glass on the left or the right? The dinner settings are symmetrical, with left indistinguishable from right, until the first diner chooses a glass and the symmetry is broken.

(c) In the symmetrical state (left), there is a unique point of minimum energy. As the universe cools down, there is no longer a preferred minimum of energy, but rather (bottom) many possibilities exist. At this time the universe is in an asymmetrical state, when, for example, there may be far more matter than antimatter.

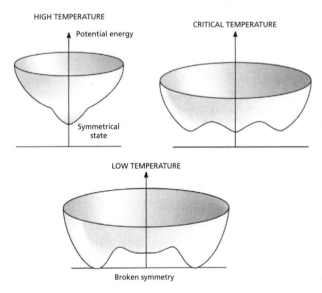

Figure 6 (a) The unification of the fundamental forces. (b) Breakdown of symmetry at a dinner table. (c) Breakdown of symmetry as the universe cools.

Source: J. Silk (1994) *A Short History of the Universe.* New York: Scientific American Library.

particles that characterized equality between the fundamental forces must have vanished. They could not be recreated as there was not enough energy available.

At high enough energy, the strong nuclear force dominates. As the temperature drops, the strong force declines. This change in the nature of the dominant force is what must have provoked a phase transition. The associated change in the state of the matter resulted in a release of energy, similar to the latent heat released when ice melts. The result was that the universe entered its second phase: inflation. We are then at 10^{-35} seconds after the Big Bang. We believe that the physics of the universe at inflation is something we can understand, although it still remains speculative for lack of any direct experimental evidence.

Inflation works by providing a constant source of energy that dominates the expansion. The energy density associated with inflation is equivalent to a negative pressure. Normally, energy associated with heat provides a positive pressure. Thermal (heat) energy cools as the universe expands. Pressure acts to do work and results in changing the energy. In a contracting gas, the effect of the pressure increases the thermal energy, as in the common phenomenon of a bicycle pump heating up as it does work by compressing gas.

Positive pressure acts like energy and mass: it is attractive. Inflationary energy is fundamentally different: like latent heat, it is constant. A constant energy density in an expanding system acts in the opposite way to thermal pressure: its effect on the expansion resembles a negative pressure. Negative pressure heats when the matter on which it acts expands. An example of negative pressure is an elastic string: expand it and the string heats up. Negative pressure is repulsive, and the universe accelerates. In the case of the universe, the negative pressure drives an acceleration that proceeds ever more rapidly.

The vacuum density boosts the expansion rate so markedly because the vacuum acts like a negative pressure. Expand the universe slightly and the pressure increases, thereby boosting the expansion further. But how did the phase transition supply the needed negative pressure boost? Imagine water freezing. It releases latent heat. The physics of fundamental interactions provides the analogous effect, when grand unification breaks down. This occurs as the universe cools below a temperature of 10^{28} K. The energy boost briefly results in exponential acceleration. The volume of the universe increased by a factor of 10^{100} or more!

It is comforting to realize that no material objects actually move at faster than light speed. Suppose that the journal *Science* doubles the number of issues published every month. Imagine inserting the new issues anywhere on the shelf. Eventually you will need a bookshelf ten light years long for the latest year's worth of issues. The ends of your collection are now moving apart at ten times the speed of light, relative to each other, but nothing is actually moving that rapidly. No material object expands faster than light. It is space that is expanding.

The energy field that drives inflation is called the inflaton. It is an energy field that has never been detected, but that physicists consider to be eminently plausible. It is a field that has no preferred direction in space. Physicists call such a field a scalar field, as opposed to a vector field, which has a preferred direction.

Inflation comes to an end when all the kinetic energy of the inflaton turns into heat. The universe has expanded by some 25 powers of 10. The universe fills with radiation. The negative pressure has vanished. Now the pressure of radiation dominates. The universe expands at its normal, relatively sedate, pace. This is the radiation-dominated phase.

From 10^{-35} seconds after the Big Bang until ten thousand years have elapsed, nothing much happens in the universe. The quantum fluctuations imprinted by inflation are present everywhere as tiny ripples in the curvature of space. These are destined to grow ever more intense until giant clouds condense out of the expanding universe under the action of their local self-gravity and form the galaxies. As long as the universe is predominantly radiation, however, no growth of these fluctuations can occur. Galaxy formation is blocked. Nothing can happen until ordinary matter dominates the universe, after ten thousand years have elapsed.

The radiation progressively fades away relative to the matter, since photons lose energy, as the universe expands until we arrive at the matter era. Matter dominates over radiation in terms of density. Now the fluctuations in density grow larger by accreting mass from their surroundings. Gravity drives the fluctuations, to become denser and denser. Eventually, the galaxies form. By now nearly a billion years has elapsed.

The final phase of creation occurs after about five billion years. The universe begins to reaccelerate under the effect of an energy field associated with the repulsive force that is known as the cosmological constant. This, like the field that drives inflation, has negative pressure, so it repels and accelerates. The energy associated with the acceleration force is dark, in the sense that there is no visible counterpart.

The dark energy is smaller than the energy field that drove inflation by about 120 factors of 10. Hence it only begins to be important when the mean density of the universe has dropped sufficiently. Then it dominates over gravitational attraction and the universe begins to accelerate. We observe this phenomenon via studies of distant supernovae, which are found to be progressively dimmer and so more distant than they would be in a non-accelerating universe.

Observations require that two-thirds of the density of the universe is in the form of dark energy. If there is a cosmological constant, its eventual domination is inevitable. The only flaw in this reasoning is that the theorists have found no compelling reason for dark energy to just now be non-zero. Observations tell us that the acceleration now dominates over gravity, and will do so for an eternity. The universe is destined to expand forever at an ever-accelerating pace.

How far back can we probe towards the beginning of time?

We can look back into the past because the universe is expanding and light has a finite speed. Look billions of light years away and we probe billions of years back in time. There we infer that the earliest phase of structure formation commenced long ago, when the universe was almost completely of uniform density. The requirements for eventual structure formation were laid down early in the expanding universe. In an expanding universe, there is not time enough for fluctuations to develop spontaneously. Rather, there must have been initial seed fluctuations that eventually gave birth to structure. The initial conditions were crucial for understanding evolution. How smooth was the early universe? If it were too uniform, galaxies would never have formed. If it were too clumpy, we would be living on the periphery of a gigantic black hole.

Theory is not much use as a guide. The breakthrough came with the observational detection of the primordial fluctuations from which structure formed. The universe is transparent. One can see far away, especially in the microwave spectral region where galaxies are dim and the primordial cosmic radiation background dominates over starlight. We know that we are able to see back to the first million years. Before then, there was an impenetrable fog: the universe was sufficiently dense that light was scattered. We must see back that far in time, given the known density of matter. But it took nearly 30 years of searching, after the cosmic microwave background radiation was discovered, in order for the temperature fluctuations to be detected.

We can even probe further back in time. We cannot see such early periods directly but the circumstantial evidence is overwhelming. This tells us of the origin of the light elements, and then even earlier there is the origin of baryons. And finally we come to the era of the origin of fluctuations, at the phase transition defined by the onset of grand unification and inflation.

We are still desperately seeking the ultimate theory of quantum gravity. Only an amalgam of quantum theory with gravity can take us to the very beginning of time. But to understand the origin of the universe, we must first refine our understanding of the ultimate nature of matter. This involves unifying gravity with the three other fundamental forces, since there is essentially no doubt that gravity, now the weakest force by far, was once on an equal footing with electromagnetism and nuclear interactions during the regime of the immense densities and pressures achieved at the beginning of the cosmos. String theory has provided indications of a possible clue to such a unifying theory.

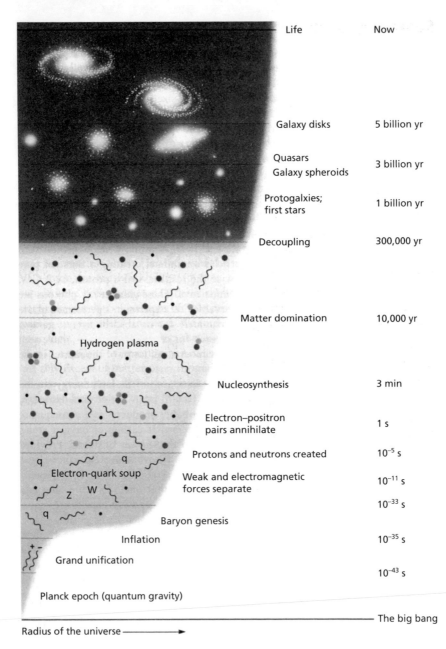

Life — Now

Galaxy disks — 5 billion yr

Quasars
Galaxy spheroids — 3 billion yr

Protogalxies;
first stars — 1 billion yr

Decoupling — 300,000 yr

Matter domination — 10,000 yr

Hydrogen plasma

Nucleosynthesis — 3 min

Electron–positron
pairs annihilate — 1 s

Protons and neutrons created — 10^{-5} s

Electron-quark soup

q   q

Weak and electromagnetic
forces separate — 10^{-11} s

Z   W — 10^{-33} s

q

Baryon genesis

Inflation — 10^{-35} s

Grand unification

10^{-43} s

Planck epoch (quantum gravity)

The big bang

Radius of the universe ⟶

Figure 7 A chronology of the universe.

Source: J. Silk (1994) *A Short History of the Universe*. New York: Scientific American Library.

Searching for the ultimate theory

> The various particles have to be taken literally as projections of a higher-dimensional reality, which cannot be accounted for in terms of any force of interaction between them.
>
> David Bohm

Superstrings are ten-dimensional objects that are subatomic particles. They are only recognizable as such once the six extraneous dimensions have collapsed in a phase transition that occurred some 10^{-43} seconds after the Big Bang. For reasons that are rather hard to explain, one can incorporate all of the properties of elementary particles, such as their masses, charges, and spins, into superstrings. The process results in a seductively simple description of both particles and the physical laws that describe their interactions. Superstrings give an elegant geometric interpretation of elementary particles. The result is a theory whose mathematical beauty is so compelling that an army of theoretical physicists has locked step to the drum beat struck by such pioneers as physicists Michael Green, John Schwarz, and Edward Witten.

Yet there is no question that the mathematics is complicated. One has to master M theory, and branes, and duality, and many other topics to enter the field. These are various ways to comprehend the universe of higher dimensions. At last count, there were some 496 different and unspecified counterparts to the photon, which in classical physics alone suffices for electromagnetism as a force carrier (the force of electromagnetism is understood in terms of the interchange of virtual photons). Little wonder that critics such as Sheldon Glashow have written:[1]

Contemplation of superstrings may evolve into an activity . . . to be conducted at schools of divinity by future equivalents of mediaeval theologians.

The hero of superstring theory, Princeton physicist Edward Witten, has rather arrogantly countered this by arguing that 'good wrong ideas that even remotely rival the majesty of string theory have never been seen'. Superstring theory has been incorporated into the theory of branes, in particular the M-branes of M theory. Any higher dimensions that once were present are today wrapped up on the scale of the Planck length and are unobservable. In the very early universe, these higher dimensions acquired reality, although they were equally unobservable. However, higher dimensions allow one the freedom to describe all of particle physics and particle interactions in terms of line-like objects called superstrings that are embedded in the universe of higher dimensions. Space–time in more than four dimensions is described by entities that are collectively called p-branes. Here p stands for any integer that represents the dimensionality of space. Physicists would very much like to predict p from first principles. We know that p cannot be one or two, for we would be respectively sausage or pancake-like, with adverse effects on our digestive systems, among other problems. Quantum gravity can be resolved, at least in principle, in spaces of higher dimensions. This has been one of the great messages from superstring theory. Troublesome infinities (and infinity really is troublesome to a physicist) can be removed if we settle for p in some higher dimensional space. The preferred number is ten, although some hold out for four. Certainly three is insufficient – one needs the extra freedom of higher dimensionality. All of this seems somehow science-fictional. Is it real? Experiments must be devised that will discriminate between these more exotic variants of cosmology.

There is no doubt that a frontal assault on the roots of cosmology is in dire need of inspiration. Cosmology so far has provided a theory of the universe that lacks a beginning. String theory may provide the answer, but its sheer complexity inevitably reminds one of Leon Lederman's dictum: 'If the basic idea is too complicated to fit on a T-shirt, it's probably wrong.'

How do we prove any of this?

One type of experiment offers some prospect of probing the first instants of the universe. Most theories of the beginning of the universe predict that a relic background of gravity waves was produced. Gravity waves are the one signal that can emerge from near the beginning of the universe. Only gravitational radiation could possibly propagate to us through the vast density of matter near the beginning of the universe, because it interacts so weakly with matter that it can traverse practically any amount of it. According to the theory of inflationary cosmology, our universe is bathed in a sea of gravity waves that were generated near the earliest instant of the universe, when quantum theory and gravity merged together. Inflation stretches out the wavelengths of these waves from the infinitesimal scale of quantum fluctuations on the Planck length. They should be present today in the universe with wavelengths of light years or more.

Gravitational radiation is measured as a change in the gravity field that propagates at the speed of light. Very sensitive combinations of test masses can measure this effect. To do cosmology this way involves searching for the traces of gravitational radiation from the beginning of the universe. Gravitational radiation is predicted by Einstein's theory of gravity. It is generated in a rapidly varying strong gravity field, such as when a black hole forms. Until now, gravitational radiation has only been detected indirectly, via the orbital precession and decay of a binary pulsar, the measurement of which won the Nobel prize for Joseph Taylor and Russell Hulse in 1993.

Gravity waves are nothing more than transient variations in the gravity field. Gravitational radiation consists of randomly oriented waves that propagate at the speed of light. Until recently, most gravity wave detectors consisted of a highly isolated bar-shaped mass that deforms slightly in response to the passage of a gravity wave. Passage of the gravity wave exerts a slight stretching force on the bar. The expected deflection is infinitesimal for known astronomical sources such as orbiting binary neutron stars.

Another approach, which ultimately leads to higher sensitivity, is to measure slight changes in the distance between two mirrors between which a laser beam is reflected many times. In effect one has a very long measuring rod that is elongated very slightly by the passage of a gravity wave because the mirror separation is slightly stretched. The sought-for effect is tiny. Several experiments are in progress using laser beams up to a kilometre long as part of an interferometer that enables a length change of a thousand trillionth a centimetre to be measured. At this level of precision, one can hope to detect the burst of gravitational radiation released when a stellar mass black hole forms anywhere in our galaxy. This is a rather rare occurrence: one might have to wait a century or more for the next event! Eventually the experimental sensitivity will be increased to the point at which one should be able to detect black hole forming events in many nearby galaxies. Various implementations of, and rivals to, the Laser Interferometer Gravitational-Wave Observatory (LIGO) are taking data, or preparing to do so, at several sites, located in the USA, Japan,

Australia, Italy, and Germany. LIGO itself is a US experiment with sites in Washington and Louisiana, and VIRGO is a Franco-Italian competitor located near Pisa.

A novel space experiment called LISA will be launched by ESA and NASA in around 2015 and will have the potential of detecting low frequency gravity waves. LISA consists of three satellites five million kilometres apart that orbit around the Sun to form a giant triangular antenna whose elements are connected by laser beams. Changes in the optical path between freely floating masses inside each of the satellites can be used to sample impinging gravitational waves that manifest themselves as infinitesimal transient disturbances in the gravity field. It is necessary to measure the separation between these satellites to a precision of a billionth of a centimetre. In fact, LISA will be able to detect the gravity wave signal emitted by the formation of a supermassive black hole weighing millions of solar masses or more that formed due to a merger between galaxies anywhere in the universe.

This experiment should detect the existence of gravitational waves. To detect an isotropic background of primordial gravity waves from near the beginning of the universe, as predicted by inflation, is far more challenging. Gravity waves are produced as a consequence of inflation, which generates all possible fluctuations in the gravity field, and an indirect imprint of the waves is left in the sky. Such a background is difficult to disentangle from instrumental noise, and will require a more sophisticated approach.

A post-LISA satellite experiment will be needed that will require a flotilla of spacecraft over separations of only 50,000 kilometres. This would measure gravity waves at unprecedented levels of sensitivity to help to discriminate the weak isotropic signal from the Big Bang. The close spacing of the satellites means good angular resolution. At these separations, close white dwarf pairs produce the gravity wave foregrounds. These are binary systems, pairs of orbiting white dwarfs. The idea is that the Big Bang Observatory will detect every close pair of white dwarfs in the universe. Once this contaminating signal is subtracted off, any signature due to inflation should be exposed. Perhaps it will show that inflation did not occur. This too would be a dramatic inference.

Of course, space interferometers are not the only means of searching for a primordial sea of gravity waves. Alternative approaches will be needed. The inflationary density perturbations survive and eventually grow and form large-scale structure. The gravity waves are only main-tained as long as the universe is dominated by radiation. During the matter-dominated period, at late times, they die away. However, we observe the cosmic microwave background radiation at about three hundred thousand years after the Big Bang, when the last interactions of radiation with matter occurred by scattering off of electrons. The gravity waves are still present and have sufficient energy to perturb the electrons that are scattering the radiation. The result is that an additional perturb-ation is imprinted on the cosmic microwave background radiation. The effect is tiny, amounting to at most one per cent of the strength of the observed fluctuations.

However, there is a potential way to identify the real gravity wave signal. Gravity waves distort the microwave background in preferred directions. These vary on the sky to result in what is called a polarized signal. Dust that is aligned by an interstellar magnetic field scatters and polarizes light, because tiny dust particles are usually elongated, and scatter the impinging electromagnetic oscillations that we call photons in a preferred direction. Atmospheric haze consists of tiny randomly oriented dust par-ticles that scatter sunlight with random polarization. Use of polarizing sunglasses captures light with a preferred polarization and so cuts down on the glare caused by scattered light.

Gravity waves produce a slight polarization of the microwave signal. The gravity wave-induced polarization has a unique characteristic that enables it to be distinguished from the more common and larger polariza-tion signal produced by scattering of the microwave background photons by electrons. Gravity waves are non-compressive: they only produce a shearing effect as they pass by, and the density is left unchanged. The slight shearing motions of the electrons produce the polarization, and differ from the symmetric compressive motions generated by density perturba-tions. The net result of the gravity waves is an antisymmetric polarization signal in the microwave background, a unique large angular scale testimony to the action of gravity waves.

A somewhat larger effect is due to the scattering by the electrons of a slightly anisotropic radiation background. But this effect is compressive, because the electrons participate in density fluctuations, and gives a distinctively symmetric polarization signal. There is a fundamental difference in the polarization produced by shear as opposed to that by compression. In principle, the two polarization signals can be distinguished because of the differing patterns on the sky. One can distinguish these signals with a detector that is sensitive to polarization. This should make the gravity wave polarization signal recognizable, when sufficiently sensitive experiments can be mounted. Cosmologists hope to achieve this goal by 2010.

Yet another approach to the relic background of gravity waves will involve using a network of the most accurate known clocks in the universe. Pulsars are rapidly rotating neutron stars, and the most rapidly spinning of these have remarkable timing stability. With a network of millisecond pulsars in different directions in the sky, a monitoring campaign over a few years would reveal correlated blips as a gravity wave passes by and simultaneously deforms space between us and several of the pulsars. This method would be sensitive to a background of gravity waves of wavelengths stretching light years across. Such ultralow frequency gravity waves are precisely what one might expect to emerge from the beginning of the universe. A new telescope, the Square Kilometre Radio Array, to be built by 2020, will monitor a hundred or more millisecond pulsars, and provide the ultimate extremely low frequency gravity wave detector. It remains to be seen how feasible such an experiment would be, as it is always possible that other systematic effects may affect the accuracy of pulsar timing. However, all indications so far indicate that millisecond pulsars are the most robust and stable clocks known to man.

The ultimate goal of detection of relic gravity waves is to probe cosmic inflation. This is our only direct means of actually 'seeing' back to near the beginning of the universe. There is no guarantee of success; indeed, some theories predict the absence of any primordial gravity waves. However, detection would be a remarkable vindication of inflation.

Note

1. These quotes are taken from T. Ferris (1997) *The Whole Shebang.* London: Weidenfeld and Nicholson.

13 The Seeds of Structure

Nature uses as little as possible of anything.

Johannes Kepler

God's first Creature, which was Light.

Francis Bacon

The primeval universe was a fireball of radiation, with the odd particle occasionally interspersed. There was no structure, and no boundary, no preferred location or direction.

The new cosmology

How did galaxies arise out of this cosmic fireball? The connection with the present-day universe can be understood by studying the structure of the universe on the largest scales. All the matter we see today was once within a sphere of radius 1 centimetre at a mere 10^{-43} of a second after the Big Bang. But how did it get so big? And why is it so uniform? Inflation, introduced in the previous chapter, provides the key.

Inflation, you will remember, is the name coined by cosmologists for a brief instant during the very early universe when the volume of space underwent an exponential increase. Inflation is the panacea for all, or almost all, of the puzzles that have plagued modern cosmology since the time of Georges Lemaître, the first physical cosmologist. Alan Guth is recognized as the founding father of inflationary cosmology. In 1982, Guth was intrigued by such questions as why the universe is so large and so uniform. He stumbled upon the answer in thinking about the possible states of matter at the beginning of time, when temperatures were so high that the fundamental forces of matter, nuclear and electromagnetic, were indistinguishable and inseparable, in fact were identical, despite being widely disparate in strength in the current, low temperature universe.

The thermal history of the Big Bang can be reconstructed back to 10^{-12} seconds. Back to this instant, we are reasonably confident of the matter and radiation content of the universe, and of the associated physics. Before this, our theoretical understanding is much more speculative.

However, the lack of a fundamental theory has never deterred cosmologists. We have seen how the theory of inflation posits that an early phase transition, occurring at the breaking of Grand Unification symmetry when the universe was barely 10^{-35} seconds old, resulted in production of a vacuum energy density that persisted for long enough to dominate the energy density of the universe. The effect on the universe is dramatic: it begins to expand exponentially rapidly. It becomes very large indeed. And like a wrinkled balloon when it is expanded, space becomes highly uniform. The vision of inflation has been developed over the past decades, and constitutes one of the most significant developments in cosmology since the discovery of the expanding universe half a century earlier.

Beginning from nothing

> The original movement, the agent, is a point that sets itself into motion. . . . A line comes into being. It goes out for a walk, so to speak, aimlessly for the sake of the walk.
>
> Paul Klee

Observational cosmologists find that to within a factor of three or so, the kinetic energy of the universe balances its gravitational potential energy. It is by no means inconceivable that the difference between a large gravitational energy and a large kinetic energy in the universe is zero. Indeed, many theoretical cosmologists believe that the zero energy state is the most natural state from the perspective of all possible initial states. Of course, physics requires that energy should be conserved, and so the universe would then have begun with zero energy. Inflationary cosmology justifies, and indeed predicts, that the universe has zero energy, but it also tells us something quite new: the universe began when both its gravitational energy and kinetic energy were arbitrarily close to zero. It literally began from nothing, or as near to nothing as makes no difference. Virtually all memory of initial conditions is erased. Exponential growth results in the ultimate free lunch.

Cosmic capitalism

The size of the universe is greatly stretched by inflation. Inflation removes wrinkles, as in an expanding balloon. But it also generates fluctuations. Just as there are inhomogeneities when water freezes, so small wrinkles remain. These are a manifestation of what physicists call a first order phase transition. These fluctuations are the seeds of structure. Quantum fluctuations become macroscopic fluctuations in density in the universe, thanks to inflation. And galaxies form from these infinitesimal fluctuations in density.

We lack a complete theory of these fluctuations. The simplest arguments suggest equal strength fluctuations, on whatever scale you examine them. This is a unique prediction and should have observable consequences, especially on large scales. One complication is that the Big Bang can act as a sort of filter that modifies the distribution of fluctuations.

Only certain types of fluctuations survive to late times. Those on small scales can be destroyed by different effects. Expansion of slightly compressed radiation is an inevitable tendency, as radiation always tends to expand and drag ordinary matter along. The result is a systematic homogenization that smooths out the smaller scale fluctuations in the matter. The limitation is only the time available since the Big Bang, so the largest scales survive unscathed.

Neutrinos, like photons, are normally considered to be massless. There is no guarantee this need be so from particle physics, and the experimental upper limits allow the possibility of neutrino dark matter. It does not take much in the way of neutrino mass to end up with a billion relic neutrinos for every proton in the universe. Indeed, a neutrino mass of a billionth of a proton mass would suffice to make neutrinos the dominant form of matter.

Particle dark matter, if characterized by massive neutrinos, has a built-in expansionist tendency. This is because neutrinos are travelling at light speed until relatively late in the life of the universe, when fluctuation growth is well under way. Neutrino dark matter is known as hot dark matter, in contrast to the very massive particles that are believed to constitute most of the dark matter. Massive particle dark matter behaves like a cold gas with negligible pressure, and is known as cold dark matter.

The largest fluctuations in hot dark matter also survive, since there has not been time enough for them to be destroyed. The neutrinos slow down only at late times because, as the temperature of the universe declines, the effect of a small neutrino mass finally becomes more important. Fluctuations on small scales are homogenized, and no seeds remain to form low mass objects. Only the large-scale fluctuations survive, eventually to form massive structures in the universe.

Most of the matter in the universe is dark. It must either be cold, clustering freely on all scales, or hot, clustering at first only on the largest scales. The resulting filter depends on the type of matter that dominates in the universe: cold or hot dark matter. Some examples of types of fluctuations that emerge from the primordial fireball, about 300,000 years after the Big Bang, are blue noise, similar to the prediction of inflation, which generates bottom-up formation, from small scales to larger scales, and red noise, which leads to a top-down sequence of structure evolution, from large scales to smaller scales.

In a bottom-up approach, small fluctuations aggregate and grow by accretion to form larger and larger fluctuations as more mass is acquired. Conversely, structure forms in a top-down sequence via fragmentation of large clouds into smaller clouds, if the large clouds are the first to form. Observations require massive objects, like galaxy clusters, to be just forming at present. Conversely, galaxies formed long ago. The theory of the evolution of galaxies tells us that many galaxies are very old systems. There is no doubt that both the data and the theory favour a bottom-up sequence for the evolution of cosmic structure. Hot dark matter can only play a very subdominant role compared to cold dark matter.

The growth of fluctuations occurs by the process of gravitational instability. A small patch of the universe is slightly overdense relative to the average patch. The gravity in this region is slightly greater, and it accretes matter from its vicinity. The process is slow: the mass doubles over the time it takes for the expanding universe itself to double in size. But the effect is inexorable. Conversely, an underdense patch loses mass as gravity slowly drains matter away. This is capitalism incarnate: the rich get richer and the poor get poorer.

All of this could well have been science fiction rather than hard science, and many remained sceptical. The paradigm shift to the Big Bang may be said to have begun in 1964, when the fossil radiation was first discovered, and to have finished by 2000, when fluctuations of a fraction of a thousandth of a per cent on the sky were found in the temperature of the background radiation. These variations traced the long sought-after seed inhomogeneities from which all large-scale structure developed. The primordial fluctuations are observed in the cosmic microwave background radiation at a level of only 1 part in 100,000.

The final pillar of the Big Bang

> A new scientific truth does not triumph by convincing its opponents and making them see the light, but rather because its opponents eventually die, and a new generation grows up that is familiar with it.
>
> Max Planck

The Big Bang rests on four rather impressive and solid pillars. Three of these have been previously discussed. The first two burst into the world with such dramatic effect that they have been called the 'golden moments' of cosmology. These were the discoveries of the expansion of the universe and of the cosmic microwave background radiation. Confirmation of the synthesis of the light elements confirmed our fiery origins.

The final pillar was the discovery of the temperature fluctuations in the cosmic microwave background radiation. The last great prediction of the Big Bang theory was that the microwave background radiation should not be completely uniform but contain fluctuations that are the fossilized seeds of galaxies and galaxy clusters. Fluctuations were first detected by the Cosmic Background Explorer satellite (COBE) in 1992. These were associated with the sought-after primordial seeds, and probed the irregularities in the universe on very large scales. Detection of similar enhancements on protocluster scales was enhanced by early fluctuation growth, a precursor to structure formation, and a prediction of inflationary cosmology. Such an enhancement on an angular scale of a degree due to the direct precursors of structure formation has now been established. The Big Bang has been definitively confirmed.

The cosmic microwave background: a brief history

Large scientific collaborations breed infighting. Sometimes this stimulates the progress of an experiment. More often, the acrimony leaves behind bitter scars. The COBE satellite, which mapped the cosmic background radiation for four years after its launch in November 1989, involved more than 1500 participants at its peak. Yet two individual scientists stand out from the crowd. John Mather was one of the prime movers behind the COBE satellite. A NASA scientist, he developed a key experiment on board the satellite, which measured the spectrum of the cosmic microwave background, the fossil radiation from the beginning of the universe. After a series of disconcerting vicissitudes that culminated in the *Challenger* Shuttle disaster, Mather oversaw the repackaging of COBE into a Delta spacecraft that made a major discovery within the first nine minutes of data taking. He targeted his experiment on the relic radiation from the Big Bang. There was a lot of history that lay behind his choice.

Most notably, we have seen how Ralph Alpher and Robert Herman, along with their mentor, cosmologist George Gamow, had predicted in the 1950s that there should be a relic radiation field permeating the universe. Alpher and Herman even came up with a temperature of 5 kelvin above absolute zero for the relic radiation of the universe. But they seem to have made no connection with microwave astronomy.

Other groups of scientists did, however. In Russia in 1964, Andre Doroshkevich and Igor Novikov made the connection between the early universe as a reservoir of blackbody radiation and microwave observations, an important step that was not made in the earlier papers by Alpher and Herman. They refer to a result obtained in 1961 by radio engineer E. A. Ohm, calibrating a satellite communications telescope at the Bell Laboratories, who reported an unexplainable source of microwave static noise, with an equivalent temperature of 3 kelvin. However, the Russian scientists lacked any new data, and interpreted the Ohm result as an upper limit, supporting a cold universe.

With hindsight, the relevant data for a hot Big Bang was already present, not just in the microwave sky but also most notably in some remarkable measurements of the gas molecule CH in interstellar medium absorption lines. In one of the most overlooked experiments of the twentieth century, Andrew McKellar discovered in 1940 that an interstellar radiation field of at least 2 kelvin was required in order to account for the molecular transitions that were being found in absorption along the lines of sight to nearby stars.

Meanwhile, in the Physics Department at Princeton University, the experimental group led by Robert Dicke rediscovered the arguments of Gamow and his associates, but in complete ignorance of the earlier work. And, more to the point, they made the connection with radio astronomy, again oblivious of the parallel Russian work. To their immense credit, they designed an experiment to search for the elusive radio glow from the beginning of the universe. Dicke set out to test the bounce model of the Big Bang, one that had a fiery past.

But Dicke and his colleagues were beaten at the final post. I have already recounted how radio astronomers Arno Penzias and Robert Wilson, taking up the Bell Laboratories experiment, although motivated by the Milky Way and not by cosmology, stumbled upon the relic fossil radiation from the Big Bang. The observable universe possessed a diffuse microwave glow, about a hundred times weaker than the static almost all of us have witnessed on a television set that is tuned between two channels. It is interesting to recall Gamow's reaction at the Texas Symposium on Relativistic Astrophysics, held in Dallas in 1967. When asked how he felt about the discovery of the cosmic microwave background, with reference to his early work, he said, 'I lost a nickel. They found a nickel. Was it my nickel?'

The theorist in the Princeton gravity group, James Peebles, had even predicted the cosmic microwave background to have the spectrum of a blackbody that would be detectable at microwave frequencies, but only published his results in 1965 after the Penzias–Wilson discovery paper appeared. The blackbody prediction turned out to be a prescient and key component of cosmology that was, however, to prove controversial for the next 25 years.

In the 1980s there were experiments, one of which was a by-product of Mather's own doctoral work at Berkeley, which purportedly demonstrated that there were unanticipated deviations from the blackbody spectrum predicted by proponents of the Big Bang theory. Cosmologists were in a quandary. The COBE team maintained absolute secrecy for two months after launch until Mather, addressing a meeting of the American Astronomical Society, showed a slide of the spectrum, with absolutely no perceptible deviations from a blackbody. The spectral accuracy surpassed that of the best laboratory blackbody. On seeing the slide, the audience spontaneously erupted into a standing ovation. Few moments in science have provided such a dramatic confrontation of prediction and observation. The press reports, however, remained muted, perhaps because a blackbody spectrum is not very familiar to the public. All changed on 23 April 1992, however, when front pages of newspapers around the world resounded with the discovery of ripples or intensity fluctuations in the cosmic microwave background radiation by the COBE team. Again, cosmologists had persevered long and hard in their search for such ripples over three previous decades, as the elusive seeds for large-scale structure in the universe. Attribution was given to a collaboration seemingly dominated by one individual, Lawrence Berkeley National Laboratory scientist George Smoot, who was primary developer of the mapping instrument on board COBE. Such accolades as Stephen Hawking's 'The greatest discovery of the century, if not of all time' accompanied banner headlines. Although the epochal discovery was soon confirmed, the sky maps of the fluctuations that appeared in the press were found in retrospect mostly to show experimental noise rather than the subtler cosmic signal. Not surprisingly, Smoot's hard-working colleagues were embittered at his moment of fame. Consternation even set in at NASA Headquarters at the lack of attribution to a heroic NASA effort that had spanned two decades. How did this public relations fiasco come about?

The efficiency of the public relations office at the Lawrence Berkeley National Laboratory, along with acquiescence in its role by Smoot, is credited with the blame. Their embargoed press release was circulated to journalists one day[1] before the press conference at which the results were announced. The journalists telephoned the world's cosmologists for comments. Years of pent-up frustration were released. Dick Bond at

Toronto and Michael Turner in Chicago simultaneously, and independently, described the fluctuations as 'the Holy Grail of Cosmology'. Joel Primack at Santa Cruz compared it to seeing 'the Handwriting of God'. The crescendo of expectation culminated in the press conference, where Smoot stated, 'if you are religious, it is like seeing God'.

But the damage was long since done. Within days, Smoot was signed up by noted literary agent and science ambulance chaser John Brockman to write the book of the COBE story along with science journalist Keay Davidson, netting them millions but spoiling the popular science book market perhaps forever. Meanwhile, a number of Smoot's former colleagues set to work busily developing a sequel to COBE, launched by NASA in the spring of 2001. The Wilkinson Microwave Anisotropy Explorer satellite has 33 times higher angular resolution and 45 times more sensitivity than COBE, and reported its first results more than a decade after COBE, in February 2003. Its results heralded an equally great success story for cosmology, which is described below.

While administrators, scientists, and engineers combined to carry off a marvellous success, one has to admit that Smoot's persistence and experimental know-how, which originally led to the first definitive measurement of the pair of hot and cold spots in the sky associated with the Earth's motion relative to the cosmic frame of the microwave background, played a key role in the fluctuation measurement. Smoot has been depicted as the jealous scientist holding up other team members from announcing the results, yet as head of the team Smoot would have been credited regardless of when the results were announced. A more plausible interpretation is that Smoot, along with the senior team member, David Wilkinson, acted their roles as sceptical arch conservatives, unwilling to let the young Turks get away with announcing preliminary results that might later turn out to need substantial modification or withdrawal. The course of cosmology is littered with such failures. COBE overcame adverse odds to be a great success.

More and more fluctuations

The density fluctuations prior to last scattering are like sound waves in a medium with a sound velocity appropriate to that of a relativistic plasma, about 70 per cent of the speed of light. After last scattering, the radiation thermally decouples and the sound speed drops to that of a gas at a few thousand kelvin. This means that the density fluctuations, which previously were pressure-driven sound waves, now respond only to gravity. Pressure is completely unimportant at least for fluctuations that contain the masses of even the smallest galaxies. Indeed, the minimum size for gravity to dominate and so for the first self-gravitating gas clouds to form is about a million solar masses. As time proceeds, the clouds build up in mass. The clouds cluster together under the action of gravity to form a galaxy and eventually a cluster of galaxies. The galaxy mass clouds are able to cool and fragment into stars. One ends up with galaxies and clusters of galaxies. The clusters contain large amounts of gas that is too hot to have cooled.

The sound waves leave a remarkable imprint on the cosmic microwave background. Inflation, or some equivalent theory, generates these waves, which just begin to undergo their first compression peak when they enter the horizon. Imagine random disturbances in this primordial fluid of radiation and baryons. These can be considered to be just like sound waves. Initially, gravity plays no role at all in affecting the strength of the sound waves. One has to wait until enough time has elapsed since the Big Bang for a given wave to undergo its first compression and begin to develop a crest. Before then, pressure has not enough time to act.

Such a wavelength, undergoing its first compression, just spans the distance travelled by light since the Big Bang. This also means that such waves would appear to have a certain angular size if one could see them on the sky. Amazingly, one can see them, as temperature hot spots or cold spots in the microwave background temperature maps. This is because there is a special moment, 300,000 years after the Big Bang, when one 'sees' the microwave radiation undergoing its last interaction with the matter by scattering off of free electrons. Why then? Because this moment is when electrons and protons combine to form atomic hydrogen, the universe being for the first time cool enough to allow this. At later instants, atoms virtually cannot scatter the microwave photons again.

Those waves, which are cresting at last scattering for the first time, have the largest amplitude. They produce a peak in the cosmic microwave background fluctuations at an angular scale that corresponds to the horizon scale at last scattering, about 1 degree. There is a competition between gravity and pressure that controls the strength of the waves. The smaller the wavelength, the greater the role of pressure in resisting compression. Shorter waves that are cresting for the second time at last scattering leave a peak on smaller angular scales, and also tend to be amplified less by gravity. Waves undergoing their first rarefaction also leave a peak on an intermediate scale, since rarefactions are also measured as temperature fluctuations. The density field is random, and fluctuations are counted whether negative or positive. There are a series of peaks predicted of decreasing strength, until one reaches wavelengths so small that they are inefficient at scattering the radiation and there are no further fluctuations. We say the primary temperature fluctuations are damped out. The damping occurs at a physical scale that corresponds to the thickness of the light travel distance during the last scattering epoch, projected on the sky. This is the distance a primordial sound wave could travel over the time the universe is undergoing the transition from ionized to neutral. This amounts to about 30,000 years, so the smallest surviving primary fluctuations are on scales of about one-tenth of a degree.

A series of peaks have been measured in the cosmic microwave background temperature fluctuations. The first, second, third, and fourth peaks have been detected. The angular position of the peaks is sensitive to the curvature of the universe. If we live, for example, in an open universe with hyperbolic geometry, the peaks are shifted to smaller angular scales. The universe acts like a giant concave lens. This effect is not seen: the universe is found to be flat to within an accuracy of a few per cent. The sum of the matter and dark energy densities adds up to the critical energy density for closing the universe.

The detection of the sound wave peaks is another independent confirmation of the dominance of non-baryonic dark matter in the universe. The peaks are produced by the inertia and self-gravity of the baryons, aided by the dark matter. The scattering of the radiation is produced by the electrons. The dark matter is important because its fluctuations are not diluted by pressure, as the radiation does not act directly on the neutral dark matter particles. It boosts the growth of density fluctuations by a factor of a few.

From the fluctuation strength, we independently infer a value of the baryon density that is about 4 per cent of the critical density. But we also require the total matter density to be about 30 per cent of the critical density in order to have enough fluctuation growth in the early universe to make the fluctuations as small as they are observed. Hence we have also an independent confirmation of the cosmological constant. For the universe to be flat, we must have a density of dark energy amounting to nearly 70 per cent of the critical density. This constitutes the concordance model of the Big Bang, one that the vast majority of cosmologists have accepted.

The concordance model received dramatic confirmation in 2003 with the announcement of new results on the cosmic microwave background fluctuations. These were obtained with the Wilkinson Microwave Anisotropy Probe, or WMAP, named after cosmic microwave background pioneer David Wilkinson. The WMAP satellite generated the first all-sky survey since COBE. With its much higher angular resolution, WMAP was able to confirm the noisier data from the earlier balloon and ground-based experiments. The first three peaks have been detected with exquisite accuracy. The underlying distribution of density fluctuations has been inferred with statistical uncertainties of a few per cent. There are unprecedentedly small uncertainties in the parameters that describe what is now the standard cosmological model. The density of dark matter and of dark energy, the rate of expansion, the curvature of the universe, the spectrum of the underlying density fluctuations: all are measured to within possible uncertainties of a few per cent.

But there was at least one unexpected result from WMAP that is keeping many cosmologists awake at night. The reionization of the universe had previously been identified with the onset of strong absorption by intergalactic clouds of atomic hydrogen in the spectra of quasars when the universe was a seventh of its present size, at about a redshift of 6. WMAP mapped the polarization of the cosmic microwave background over a wide range of angular scales. First, the good news that was anticipated. Scattering by electrons is responsible for the polarization, and was found as expected to correspond to the last scatterings that occurred at a redshift of about 1000. This translates to an angular scale of around a degree.

The polarization peak is actually at a larger angular scale than the principal peak in the temperature fluctuations. This tells us that there were fluctuations on scales larger than the horizon of the universe when last scattering occurred. This is observational verification of superhorizon structure, a prediction of inflation.

The fluctuations must have been created by superhorizon physics and in particular the presence of fluctuations created by inflation. Superhorizon means that light communication could not have occurred across these scales between the time of the Big Bang and last scattering, in a Friedmann universe. This is evidence for inflation. A unique consequence of inflation is the superhorizon or acausal physics that imprints fluctuations on scales that move outside the horizon as the universe inflates. Only much later do the fluctuations re-enter the horizon.

Now for the unexpected news. An increase in the polarization was also found at very large angular scales. This was predicted to be the signature of the reionization of the universe. The polarization results from scattering of the cosmic microwave background by the newly created electrons. However, it was anticipated from the quasar evidence to occur when the universe was a seventh of its present size, at redshift 6. In fact, more late scattering was inferred than expected. The extra scattering corresponds to reionization when the universe was one-eighteenth of its present size. One needs the higher density of the universe at earlier times to give the extra scatterings.

This represents a challenge to conventional models in which the reionization is simply produced by the first massive stars. There are serious doubts as to whether one has an adequate supply of massive stars so early in the standard model. No consensus has been reached. One has to admit that it helps to liven up the field of cosmology for not everything to fall into place. There are few dull moments for a modern cosmologist, nor can she afford to take her eyes off the data for very long.

The rise of structure

The universe, then, was nearly, but not completely, homogeneous. Density fluctuations were present on all scales, as a consequence of inflation. But nothing much happened for the first ten thousand years because there was too much radiation: indeed, the density of matter was mostly in the form of radiation. This made it virtually impossible for matter to condense; it was simply too hot.

After ten thousand years elapsed, the radiation had degraded in energy and mass relative to the matter because of the expansion. The mass in matter does not change within any closed boundary that expands freely with the universe. But the radiation content within this region does change. This is because photons, which are pure energy, lose energy and so lose mass. Photons lose energy in proportion to the amount by which the volume expands. So the net amount of energy, and the equivalent amount of mass inferred from Einstein's famous equation $E = mc^2$, requires the mass density in radiation to decrease relative to that in the matter as the universe expands. We say that photons are redshifted, while protons are not. Matter comes to dominate the density, and gravity can then operate.

Galaxies formed first at the extreme, and hence rare, density peaks in the primordial density field. Even if the underlying fluctuations are completely random, rare peaks will be clustered together. An analogy is a mountain range such as the Himalayas, where the highest mountains occur together. The average peaks, which resemble hills rather than mountains, are more uniformly distributed. Most of the matter is outside the concentrations of galaxies, rather than centred on them. Most of space is filled by large regions, containing only average peaks, and devoid of early forming galaxies. Just as mountain chains occur, occasional groupings of peaks are found that are filamentary or sheet-like in shape.

Interactions between galaxy-mass clouds aid star formation by compressing and concentrating the gas supply. Galaxy formation is triggered by encounters and mergers between the clouds, and such encounters are common in the early universe. In the late universe, galaxies are so far apart that mergers are rare. In this way, galaxy formation is biased towards the high peaks, the filaments, ridges, sheets, and knots where the density is locally enhanced. Surrounding the galaxies, which aggregate into clusters and groups, are large regions that are mostly devoid of luminous galaxies.

Structure formation can be followed on large computers by the use of numerical simulations. Take a cloud of mass points, which can be treated like billiard balls that never collide, because they are individually so small and far apart, but exert forces among themselves via their gravitational attraction. A computer is needed to follow enough particles, many millions being needed in practice, to simulate the formation of a galaxy under

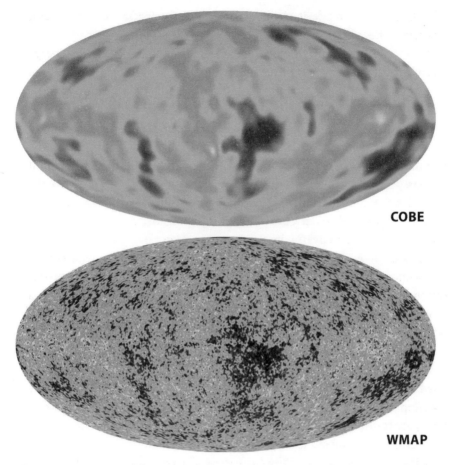

COBE

WMAP

Figure 8 Anisotropy of the cosmic microwave background radiation as seen by the COBE (1992) and WMAP (2003) satellites.

Source: NASA.

gravity. A simple equation describes the force between each pair of particles. If there are N particles, there are N^2 possible combinations of particle pairs, and N^2 equations to be solved. With the aid of a super-computer, the gravitational clumping of the particles can be studied for N as large as a hundred million. Even this number may not suffice: after all, there are about a hundred billion stars in the Milky Way galaxy, as well as thousands of gas clouds. Nevertheless, such a large simulation should provide a good approximation to a galaxy.

Dark matter is needed to provide enough gravity if inflation provided the initial conditions. Particles of cold dark matter are not laid down at random. Gravity is at work and results in an innate tendency for fluctuations to be clustered, or correlated. Cold dark matter fits the clumpiness that is observed.

As an expanding cloud of cold collisionless particles evolves under gravity, small fluctuations grow stronger and merge into larger fluctuations. The evolution is bottom-up. Dark haloes form in this manner and stay diffuse, since the weakly interacting particles that constitute the dark matter cannot lose any energy and thereby collapse further. However, the baryons associated with the dark matter, amounting to about 5 per cent of the mass, are able to radiate and concentrate into clouds of gas embedded within more diffuse dark haloes.

Cosmology progresses by eliminating alternative models. The most striking success of this philosophy is the death of the hot dark matter theory as a top-down scenario for structure formation. We have seen how hot dark matter, in which the collisionless particles are too hot to clump on galaxy scales, requires clusters to form first. The ensuing evolution is top-down. This does not agree with what we observe. In a hot dark matter dominated universe, the typical structures are cluster mass, whereas in a cold dark matter dominated universe, galaxies and even smaller systems are predominant. We observe a universe that resembles the cold dark matter dominated situation. If galaxies form by mergers of smaller clumps, in a bottom-up sequence, an acceptable match is obtained to the large three-dimensional surveys of the galaxy distribution.

The merging results in the formation of galaxy haloes. We can understand the extent of dark haloes and their masses. Dark haloes are predicted to be rotating, but only very slowly. Torquing forces are induced by gravity between neighbouring fluctuations in the expanding universe. The rotation arises as the gas cools and concentrates within the dark halo. Angular momentum is conserved, and the gas begins to rotate more rapidly. In this way, a rotating disc is eventually formed. Any further contraction is prevented by rotational support. The next stage of the galaxy formation process remains a mystery. To form the luminous regions of galaxies is a complex process. The numerical modeller must acquire mastery of gas physics and the complexities of star formation. We are still far from this goal.

Filaments and sheets

There are some very large agglomerations of galaxies. These are identified as clusters and superclusters of galaxies. Clusters of galaxies reveal themselves easily. There may be a thousand galaxies or more within a cluster, all within a volume spanning a few million light years. The galaxies move on randomly oriented closed orbits, always confined to the same region. A galaxy cluster is a self-gravitating ensemble, recognizable in distinct ways. One can count the galaxies in an optical image and see a great concentration. The centre-of-mass motion of the cluster follows the Hubble expansion. Relative to the cluster center, the galaxies have high space velocities as measured from their spectrum by the Doppler effect. The local gravity of the cluster is strong enough for the concentration of galaxies to survive, and indeed to grow as more galaxies are accreted.

X-ray images of a galaxy cluster show a diffuse glow. This is X-radiation produced by the intergalactic gas heated to tens of millions of degrees. The gas is hot because of the heating induced by the collapse of the galaxies and gas to form the cluster. We have already seen how the pressure of the gas acts against the gravity force, and provides an independent measure of the cluster mass. The energy distribution of photons in the microwave background radiation is distorted on passing through the hot gas. At low frequencies there is a slight deficit of photons, while at high frequencies there is a slight excess. This is because photons gain a small amount of energy by scattering off the hot electrons in the intracluster gas. A radio map of the cluster at a wavelength of a few centimetres shows an apparent hole in the cosmic microwave background, in reality a cold spot. At sub-millimetre wavelengths, however, one sees an excess of microwave radiation, in effect a hot spot. The gas pressure, the product of electron density and temperature, can be measured by this distortion.

Finally, the images of background galaxies are distorted when viewed through the cluster by its gravity field. This is a manifestation of the phenomenon of the bending of light rays by gravity, which results in the gravitational lensing of background images. Gravitational lensing maps of clusters provide a powerful probe of the dark matter distribution. All of these techniques are used to study clusters of galaxies and to infer their masses. About a thousand galaxy clusters have been counted out to a distance of one and a half billion light years, or out to 10 per cent of the observable universe. Beyond this distance, the data become incomplete. A typical great cluster weighs in at about a thousand trillion solar masses. By way of comparison, the dark halo around our Milky Way has a mass of a trillion solar masses.

Outside and connected to the great clusters are great sheets and filaments of galaxies, so that the universe resembles a complex web defined by the galaxies. Clusters are found at the intersections of sheets and filaments. Occasionally, there are groupings of clusters. These are known as superclusters, and may contain up to ten or more clusters. About 30 superclusters have been identified, again to within about a tenth of the visible horizon. These are the largest entities of matter in the universe.

The inevitability of cooling

Once gas clouds of galactic mass form, they undergo collapse, because the attractive force of gravity dominates over the dispersive tendency of gas pressure. Unlike a cloud of stars, a cloud of gas loses thermal energy. Atoms collide and such collisions cause electrons to jump to higher energy levels. The atoms de-excite as the electrons decay into the ground levels of the atoms by emitting photons. In this way a gas cloud cools. The result is that the gas pressure cannot counter gravity, and under the action of its own gravity, the gas cloud becomes denser and denser. It may break up into smaller cold gas clumps, as the gas pressure is further reduced because of the cooling. The gas eventually becomes so dense that the radiation is absorbed before it escapes from the cloud. At this point, the opaque clumps that have formed are well on the way to being stars.

The gas cloud fragments into dense pressure-supported stellar-mass clumps of cold gas, which eventually form stars. Once stars are formed, the evolution of the stellar system is frozen. This is because, unlike a gas cloud, a cloud of stars consists of objects that do not collide with each other, as stars are relatively compact. No energy can be lost from the system once a galaxy has formed. If the star formation occurs early and is followed by a series of mergers between smaller clouds of stars, the natural outcome is a galactic spheroid or an elliptical galaxy. If a lot of gas is present and star formation occurs late, the natural outcome is a disc-like system, because a massive gas cloud tends to flatten along the axis of rotation.

The first episode of star formation

Star formation is a complex process. It is poorly understood, despite the fact that we observe star-forming regions in great detail. There is, as yet, no fundamental understanding of how stars form. We cannot predict the mass distribution of the stars. Massive stars form heavy elements and explode, recycling most of their mass. Less massive stars contribute most of the light in the universe, and the least massive stars account for most of the mass of stellar systems. Hence we need some understanding of stellar origins if we are to make progress towards understanding how galaxies form.

There is some good news, however. A primordial cloud is much simpler to model than a typical star-forming cloud. In the Orion star-forming region, a few hundred parsecs away, stars are observed to be forming in dusty, magnetized, molecule-rich environments. The physics of star formation involves the fragmentation of a cooling, collapsing cloud. It is compounded by such additional complications as the effects of magnetic fields in providing pressure support, by the coagulation of cloud clumps driven by turbulence and accretion, and by the role of complex interstellar chemistry in regulating the rate at which a collapsing cloud can lose energy. Each of these issues merits an entire volume. Even this would merely suffice to scratch the surface of the problem, and not solve it. Studying the fate of a nearby interstellar cloud is more complex than long-term weather predictions, another task that involves mastering the fate of atmospheric clouds, and that is beyond the scope of the largest supercomputers.

A primeval cloud offers certain advantages. There are no magnetic fields, there is no dust to complicate the chemistry, and there are no complex molecules or heavy atoms. Before any stars have formed, the physics is relatively simple. Without such complications, the physics of cloud collapse can be computed numerically. The results are quite different from what one might expect by analogy with nearby clouds.

There is one aspect in common with all star-forming clouds. The initial clouds themselves contain about a hundred thousand solar masses of gas. But that is where the similarity ends. In primordial conditions, as will be described below, only very massive stars, amounting to about a hundred solar masses or more, seem to form. The cooling is so inefficient that smaller mass gas clumps do not fragment out. Such massive stars will be short-lived. Within a million years, a one hundred solar mass star will have exhausted its nuclear fuel supply and exploded as a supernova. Its debris will enrich the environment, which will be subsequently incorporated into more massive galaxies. Primordial star formation is an environmentally friendly process, in that a central role is played by recycling of stellar debris into new generations of stars.

Discs form

All galaxies rotate to some extent. Rotational forces support disc galaxies. The rotation is induced by the gravitational pull between neighbouring clouds of matter in the early universe. The clouds are irregular in shape. One side of a cloud experiences a greater gravitational attraction to its neighbour than the far side. The consequence is that a tumbling motion develops. The gas cools and becomes denser, and the clump contracts. The gas spins more rapidly as it does so. Soon it resembles a disc-like structure, supported at the periphery by rotation. The gas disc eventually breaks up into clumps, which in turn fragment into the first generation of stars.

Cold discs of gas are very unstable objects. The gas inevitably cools, and a thin disc develops. The disc will form a large bar-shaped concentration and break up into smaller aggregations of gas, which in turn will form stars. The presence of a dark halo slows down this process, preventing the bar from forming. The dark halo also provides a reservoir of gas that can continue to fall into the disc and provide the raw material that continues to form stars. Without gas infall, many spiral galaxies would long ago have exhausted their gas reservoir. We would identify them as S0 galaxies, inert systems containing exclusively old stars. Thanks to gas infall, one can understand a remarkable property of galaxies like our own. Disc galaxies retain a youthful appearance. Stars form from the gas and the massive stars explode, ejecting heavy elements such as iron and carbon into the interstellar medium. New metal-rich stars form out of the enriched gas. The elemental abundances of stars therefore provide a crude measure of age: metal-poor stars are old, metal-rich stars are younger.

There are relatively few metal-poor stars. If our galaxy formed initially as a closed system with no late addition of primitive gas, the early generations of stars could have formed when relatively few supernovae had exploded. The enrichment of the gas from which they condensed would have been minimal. There would today be a substantial population of long-lived metal-poor stars from this early epoch. However, if infall occurs of relatively pristine gas, most of the galaxy disc arrives late, after there have already been a substantial number of supernovae. The net effect is that the number of extremely metal-poor stars is diluted. We do indeed observe that the typical abundance of metals in disc stars is not very low. However, only 1 per cent of the solar value would be expected in a closed system. Instead, the disc is found to be at about a third of the solar value. This is strong indirect evidence that substantial amounts of gas infall occurred on to the forming disc. This infalling gas must have had at most a low abundance of heavy elements, not exceeding more than about a tenth of the solar value. This is very similar to what we observe in intergalactic gas clouds, which are most likely to provide the reservoir of cold metal-poor gas that refuels the star-forming disc galaxies.

Elliptical galaxies

Approximately two-thirds of all the stars in the universe are not in disc galaxies but in elliptical galaxies. Discs, being brighter by virtue of the more massive stars they contain, dominate the light. However, spheroids dominate the stellar mass. Ellipticals are pure spheroids, whereas disc galaxies are a hybrid of disc and spheroid. We understand why a disc breaks up into clouds and then stars: a cold disc is gravitationally unstable. How a spheroid of stars forms is less obvious. There are two possibilities, both of which are thought to play a role. A disc of stars, once the gas is mostly exhausted, can only heat up. Stellar collisions occur rarely, if at all. There is little gas left to radiate away energy. A stellar bulge, and occasionally a bar, develops in the stellar distribution.

Such a slow evolutionary process, which takes many disc rotation times, can plausibly account for small spheroids, such as that of the Milky Way. Our spheroid contains perhaps 10 per cent of the mass of the disc. But a more dramatic process must have occurred for massive spheroids. A merger between two disc galaxies results in the formation of an elliptical galaxy. The stellar orbits mix up dynamically, and the system settles down after a single dynamical time into a massive spheroid. Dynamical relaxation is rapid in a situation where the gravity field is undergoing violent changes. Any residual gas may settle later into a disc, but this is likely to be a subdominant component of the final system.

The first stars

As we have argued, the first stars were probably massive and so short-lived. We believe this not because we have any direct evidence, but from theoretical reasoning and indirect traces of long-dead stars. The underlying problem, as mentioned before, is that we have no fundamental theory of star formation. We cannot explain the masses of stars that have recently formed in nearby molecular clouds, where there are extensive observations. But modelling a primeval cloud is rather simpler. Once stars form, questions of feedback arise, which greatly complicate the ensuing evolution. However, the challenge of evaluating the characteristic mass of the first stars to form is mostly a numerical challenge. It involves mastering the enormous dynamic range between that of a galaxy and that of a star, or, at least, that of the parent gas cloud from which the star condensed. This can be done. The masses of the first stellar mass clumps have been evaluated by numerical simulation of the collapse of a cloud of primitive composition. The cloud contained a million solar masses of gas, characteristic of the first baryon clouds in the universe. Such a cloud is a typical building block for a galaxy. At an epoch of a billion years after the Big Bang, these clouds represented the scale of typical structures in the universe. The densest patches in the cloud form molecular hydrogen. Not much is formed – at most a tenth of a per cent of the hydrogen atoms combine into molecules – but this suffices to trigger cooling. No dust is needed to catalyse molecule formation, as happens in the nearby interstellar medium. The hydrogen molecules form by simple ion chemistry involving atoms of hydrogen to which free electrons occasionally stick to form a simple negatively charged hydrogen ion. These in turn acquire protons to form hydrogen molecules, H_2.

The formation of hydrogen molecules allows cooling to occur, because molecules of hydrogen have lower internal energy levels associated with their rotation than do electrons in atoms. The densest clumps are able to cool to about 1000 kelvin, at which point they are dense enough to fragment. The fragments continue to cool and get denser as they are able to contract further. More break up occurs, until clumps containing about a hundred solar masses form. This appears to be the end of the road for fragmentation, and no more occurs.

A dense core of a hundredth of a solar mass forms in the centre of the gas clump, as a star begins to form. The core accretes surrounding gas and grows in mass. At 1000 kelvin, the accretion is relatively rapid, amounting to a thousandth of a solar mass of gas per year being accreted into the forming star. By way of contrast, a similar calculation in nearby molecular cloud clumps comes up with a very different conclusion, because the molecular clumps in nearby dense clouds are much colder. A typical temperature is only 10 kelvin, from which we deduce an accretion rate that is a thousand times smaller than for a primordial gas clump. Nearby accretion is slow. As a result, today stars of a wide range of masses are formed, from a tenth of a solar mass to a hundred solar masses. We infer, however, that in primordial clouds only massive stars would have formed in view of the high accretion rates. Most of the gas clump is accreted, and we conclude that primordial stars are generally massive, weighing in at thirty to a hundred solar masses. Some of the first stars may have weighed up to a thousand solar masses.

A star as massive as this has a lifetime of no more than a million years. It ends its thermonuclear life by exploding and liberating large amounts of debris from its core. The debris has been enriched by nuclear reactions and includes elements such as oxygen, sulphur, and iron. The explosion may leave a black hole of a few solar masses behind, or may even be completely catastrophic in a narrow mass range around two hundred solar masses. Regardless of its detailed fate, however, the pristine hydrogen of the surrounding protogalaxy is contaminated by heavy elements. In fact, depending on the details of the death of the massive star, there will be differing imprints of its nucleosynthetic signature on the surrounding gas.

A massive star that leaves a black hole behind produces less extreme ratios of nuclei with odd and even atomic numbers than one which disrupts completely. Just as a single point of support for a weight is less stable than two points of support, one finds that even nuclei are generally more stable than odd nuclei, and so are more abundant in nature. In fact, the details of the thermonuclear explosion will lead to a preponderance of even over odd nuclei, but the ratios depend in detail on the type of explosion, and on the mass and history of the exploding star. The ratios of even to odd nuclei probe the detailed thermonuclear history of the explosion. The next generation of stars to form out of the surrounding gas will display the nucleosynthetic signature of the first stars. This leads to an indirect test of the inference that primordial stars were massive, since some of these second generation stars will have been of low enough mass to have survived until today. A unique nucleosynthetic signature is generated by a star of 100 solar masses. Of course in practice there will have been a range of masses, and this will complicate the predictions.

One has to search carefully for traces of the first stars. Our galactic halo is the logical hunting ground. This is where the first galactic building blocks were assembled. As the gas eventually accreted to form the disc, stars were left behind to orbit the galaxy forever. These stars were formed out of the gas contaminated by the short-lived first stars. The second-generation stars would have spanned a full range of masses. It is the low mass survivors that we might hope to detect today.

Such stars would be members of the galactic halo population and distinguishable by both their low chemical abundances and high radial velocities and proper motions. We seek the oldest stellar survivors by surveying the halo for candidate stars with a fraction of metals such as iron that is a ten-thousandth or less than its value in the Sun. Successive generations of stars are progressively enriched, culminating eventually in solar abundances that were attained five billion years ago in the disc of the Milky Way, when the Sun formed. The oldest stars in the galaxy formed about 12 billion years ago, ample time for several generations of stars.

Two stars have been discovered to have an iron abundance that is only a hundred-thousandth of that of the Sun. These are the record holders for the most primitive known stars. The abundances of the various heavy elements seen in the spectrum of these stars, as also is the case for less extreme stars with a ten-thousandth or even a thousandth of the metallicity of the Sun, require a massive, long dead, progenitor star.

This is in contrast to what one finds with stars that are more enriched, to above, say, 1 per cent of the solar abundance. Such stars have abundance ratios that can be accounted for by a normal mix of stars, along with their concomitant supernovae. Such a parent population resembles the stars one sees forming and dying in nearby regions of star formation. A long-extinct generation of very massive stars is required to account for the chemical abundance signatures of the most primitive, metal-poor stars in the galaxy. Those stars, of which we only have a fossil record, were the first stars.

The first galaxies

Simulations (including gas) of merging galaxies show that the transient stages of mergers can look like very irregular galaxies. We see the transient development of tidal tails and spiral arms. But beyond these transient patterns associated with ongoing star formation, there are symmetrical systems, such as discs and spheroids formed from older stars. Studies of the morphology of galaxies enable us to probe the initial conditions, because the current timescale for change would be far longer than the age of the universe. Galaxy shapes must have arisen during their formation phase. For a galaxy to be round, its stars must have formed relatively rapidly, over the time for a merger to occur. If we were to wait much longer, the gas would slowly spiral and fragment into a disc of stars. We conclude from this that the birth of ellipticals is likely to have been an ultraluminous event, whereas disc galaxies form in a more sedentary fashion, characterized by a lower star formation rate and luminosity.

Compare these speculations with the rare examples of nearby merging galaxies. Mergers are an inevitable outcome as the hierarchy of structure develops. They were common long ago. Only the most pathological cases remain today, but these are often the most interesting. A merger produces a luminous radio galaxy or a strong burst of star formation. Objects that otherwise might appear to be 'normal' galaxies are found on closer inspection to be undergoing a recent merger that is fuelling a strong burst of star formation.

Mergers in the past leave their traces in nearby galaxies. Faint shells are detected in deep exposures of some elliptical galaxies. These are like the ripples remaining on the surface of a pond, long after the stone has sunk: fossils of a merger that occurred eons ago, more than two billion years in the past.

Mergers should form elliptical galaxies. Not surprisingly, the oldest galaxies are expected to be the ellipticals. These formed long ago, but at redshifts that are accessible to modern telescopes. Ultraluminous galaxies have indeed been discovered in the distant universe. Most of their light is produced at infrared wavelengths. The rate at which they are forming stars is prodigious, amounting to a thousand times that of the Milky Way. Close inspection of the images reveals that there is almost invariably an ongoing merger between galaxies. It is natural during such a merger that the interstellar gas and associated dust is highly concentrated. The fireworks of star formation are obscured, but the light must escape at infrared wavelengths, where it is re-emitted by the dust. These so-called ULIRGs (ultraluminous infrared galaxies) are excellent candidates for forming ellipticals.

Young star-forming disc galaxies have been discovered by a novel technique using optical wavebands that searches the universe for star-forming galaxies at high redshift. Use of filters that look for an abrupt fall-off in the light from a galaxy below a specified wavelength can search for the so-called Lyman break. This is the effect of the absorption by hydrogen atoms of far ultraviolet radiation. It occurs in all gas-rich galaxies, at a wavelength of 91.2 nanometres. By searching for the Lyman break at much longer wavelength in a deep image, one can thereby select a sample of highly redshifted galaxies. In this way, large numbers of star-forming galaxies have been selected at distances of 10 megaparsecs or more, corresponding to redshifts of up to 4 when the universe was a fifth of its present size and the Lyman break was at 456 nanometres. These galaxies are the youthful counterparts of Milky Way-type galaxies.

Armed with complete samples of star-forming galaxies that are randomly selected in specific areas of the sky between the present epoch and high redshift, one can now simply add up the luminosity observed in a given volume of the universe. Luminosity translates directly into the rate at which stars are forming, since it is the massive, short-lived stars that dominate the luminosity. The result is that star formation appears to rise rapidly as one goes into the past, and peaks when the universe was a third of its present size. Star formation continues to when the universe was a quarter of its present size, at a redshift of at least 5. Hence most stars were formed when the universe was about one-third of its present age. Star formation has been declining ever since.

The distant galaxies are generally compact, are forming stars, and are spatially clustered. The compactness suggests that they are elliptical galaxies or spheroidal components of disc galaxies in formation. The spheroids are the oldest components of disc galaxies. The clustering confirms the expectation from the bottom-up theory of structure formation that the overdense regions of the universe have an early start relative to the underdense regions.

Our Sun: from past to future

The Milky Way galaxy condensed from a massive gas cloud some 12 billion years ago. The gas possessed some angular momentum, acquired as a consequence of tidal torques between neighbouring protogalactic clouds that were drifting apart in the expanding universe. As the gas contracted, conservation of angular momentum resulted in the main cloud spinning more rapidly, and the disc of the Milky Way formed. The disc broke up into many smaller gas clouds that in turn aggregated as they orbited the disc. Eventually massive cloud complexes developed and fragmented into stars.

Our Sun was born 4.6 billion years ago in a dense cloud of interstellar gas and dust. The cloud had steadily accumulated interstellar matter as it orbited the Milky Way for some hundreds of millions of years. The gas composition was far from pristine: previous generations of dying stars enriched the heavy element content of the gas. About 2 per cent of the gas was in heavy elements such as carbon, oxygen, and iron. Consequently the gas was able to cool by radiation in various atomic and molecular transitions.

The most common molecule in the interstellar medium, after hydrogen, is carbon monoxide, and this is the most effective of the molecular coolants. Astronomers map out its emission at microwave frequencies and find that the Milky Way contains thousands of molecular clouds. All of these are sites, or potential sites, of star formation. As the clouds build up mass by coalescing with other clouds, the force of self-gravity eventually dominates over the gradient of thermal pressure. The clouds contract. Magnetic forces play a role in inhibiting the cloud from collapse. Clouds of a few thousand solar masses contract slowly, but in much more massive clouds, of hundreds of thousands of solar masses or more, gravitational collapse inevitably occurs.

Even here, though, the collapse is slowed down. Once stars start forming, there is energetic feedback via outflows from young stars that heats the cloud and imparts momentum to the molecular gas. The net effect is that star formation is an inefficient process. The collapse time of a cloud is a million years, yet a cloud typically continues to form stars over tens of millions of years. The Milky Way still contains many molecular clouds that are disrupted once star formation has exhausted even a few per cent of their mass.

Dying stars explode and are the source of cloud disruption. These are massive stars, of more than ten solar masses, that end up as supernovae. The Sun itself will have a more quiescent fate. The fate of a star is determined primarily by its mass. The future of the Sun is predestined, and its past and future hold few mysteries. The evolution of stars is driven by the precarious balance between self-gravity pulling inwards and thermal pressure exerted outwards. Energy lost by radiation must be balanced by that produced from thermonuclear reactions in the core. As fuel is exhausted, the star must contract.

Nuclear reactions commence once the core of the protosun has reached a temperature of about a million kelvin. The first reactions involve burning of deuterium, an isotope of hydrogen. About one-hundredth of a per cent of the hydrogen is in deuterium. This is a sufficiently small fraction of the mass that the consequences are to delay the continuing of contraction of the protosun rather than to reach a steady state of deuterium burning. Eventually the central temperature reaches ten million kelvin. At this point the temperature is high enough to overcome the repulsion between positively charged protons. Thermonuclear fusion of helium to hydrogen occurs in a reaction cycle, which can be summarized as four hydrogen nuclei forming one helium nucleus. The helium nucleus of atomic mass 4 consists of two protons and two neutrons. It weighs fractionally, by seven parts in a thousand, less than the few protons, which initiate the reaction chain. The energy equivalent of this mass difference is generated. The net effect is energy release. The Sun will continue to survive on fuel from its hydrogen core for about another five billion years.

The hot core amounts to about 10 per cent of the mass of the Sun. The mass of the Sun is 2×10^{33} grams, so that the energy reserve of the core is 0.7 per cent of the mass of hydrogen in the stellar core. The energy equivalent amounts to 1.2×10^{44} joules. For comparison, the world reserve of fossil fuel would provide enough energy to sustain a rate of 1 gigawatt for a few hundred years, or about 10^{19} joules, an infinitesimal fraction of the solar supply.

The Sun is a prolific user of energy and produces some four hundred thousand trillion gigawatts of power, radiating 4×10^{26} joules every second into space. The Sun produces energy equivalent to a million trillion commercial nuclear reactors. It will continue to radiate at this rate for five billion years before the solar supply of nuclear fuel is exhausted. Hence the Sun is a middle-aged star, half way along to hydrogen core exhaustion. The thermonuclear energy supply allows it to maintain the same size and temperature, with gravity balancing pressure as long as the fuel supply maintains the luminosity of the Sun. The Sun is pressurized by the core energy release because it is an opaque ball of gas: the energetic particles and gamma rays produced in the nuclear reactions get degraded and eventually, after many absorptions and emissions, provide the innocuous source of yellow light from the surface of the Sun.

The fate of the Sun

Once the hydrogen in the core is exhausted, the source of heat and pressure is removed. Self-gravity dominates, and the core contracts. The result of core contraction, however, is that the core heats up. Soon it is at a temperature of 100 million kelvin. At this temperature helium undergoes nuclear reactions to form carbon via the triple alpha process: three helium nuclei, or alpha particles, fuse into a single carbon nucleus. The Sun has a new lease of life. Moreover, it is a dramatic new phase for the helium core, which is itself surrounded by a larger hydrogen burning core that is up to ten thousand times more luminous than the Sun was during its pure hydrogen burning phase.

The net result is that the pressure enhancement associated with the sudden increase in luminosity by the core destabilizes the outer region of the Sun. The outer envelope of the Sun expands. The effective temperature decreases, and the light reddens. The Sun becomes a red giant. The outer envelope expands beyond the orbit of Jupiter. The Earth is incinerated. Five billion years into the future, we may hope that the Earth's inhabitants will have found a solution for escape from the fiery fate of the home planet.

As the envelope expands further, the outer layers are puffed out into space, and the Sun turns into what astronomers recognize as a planetary nebula. In the centre of a planetary nebula a bright, white-hot star is usually visible. This is destined to cool into the ultimate relic of the solar core, a white dwarf star. A white dwarf consists of a mixture of helium, carbon, and oxygen. The mass of the precursor star helps to determine how far the core evolved along the thermonuclear evolution scheme, where hydrogen burns to helium, helium to carbon, and carbon to oxygen.

The end point of thermonuclear evolution is determined by degeneracy. This state of matter occurs when the density is so high that a novel type of pressure associated with the quantum theory becomes more important than thermal pressure of the hot electrons. This new form of pressure is associated with Heisenberg's uncertainty principle. According to this principle, there is an inevitable uncertainty associated with the position of an elementary particle. This uncertainty is equivalent to a pressure, but it is only significant compared to thermal particle motions at extremely high density. Once the quantum pressure is important, however, the star reaches a stable state of equilibrium in which thermonuclear energy no longer plays any role.

We have seen that there are two types of particles that constitute matter: light particles like electrons, called leptons, and heavy particles like protons, called hadrons. Both are collectively referred to as baryons. When atoms are squeezed together under such high pressure that the atoms are virtually overlapping, some of the electrons remain more

mobile, in particular due to the quantum uncertainties in their locations. This is what gives rise to the phenomenon of quantum pressure. In
this case we say that the electrons are degenerate, not belonging to any
particular atom.

Electron degeneracy pressure becomes important when the Sun has
shrunk by a factor of about 100 relative to its present size. The density will
then have reached a value of 1 tonne per cubic centimeter. The mean
separation between particles is then approximately the Compton wavelength of an electron. The Compton wavelength of a particle determines
the scale when its quantum properties, or its wave-like behaviour,
becomes dominant. Electrons cannot be packed any more tightly together:
this is the source of the degeneracy pressure. A white dwarf has a radius of
about 1000 kilometres and shines only via its residual thermal energy. It
forms as a hot object at the end of the thermonuclear reaction phase, and
subsequently fades away to a black dwarf as it cools down after tens of
millions of years. The Sun will turn into a white dwarf, surrounded by a
handful of surviving planets beyond the orbit of Mars. The maximum
mass of a white dwarf is 1.4 solar masses. Stars up to eight solar masses
end up as white dwarfs after shedding most of their mass in the planetary
nebula phase.

The fate of a massive star

A star of more than ten times the mass of the Sun dies a more violent
death. It collapses once it runs out of thermonuclear fuel. Its self-gravity is
high enough to guarantee thermonuclear fusion all the way to iron. Iron is
the end point, however, for release of thermonuclear energy by fusion.
Fusion of iron nuclei cannot release any further energy. The only elements
more massive than iron that release nuclear energy are the unstable uran-
ium-like isotopes that undergo nuclear fission. The star ends up with a
compact iron core that lacks a fuel supply and must collapse. Degeneracy
again comes to the rescue, but only after the iron has decomposed into
neutrons and protons, and the protons turn into neutrons via capture of
an electron and emission of a neutrino. The neutrinos are weakly interacting particles that are virtually massless, and pass almost unimpeded

through the outer layers of the collapsing core. A burst of neutrinos was measured in 1987 in three deep underground neutrino detectors in different continents when a supernova exploded in our nearest neighbour galaxy, the Large Magellanic Cloud. The star now has a dense core of neutrons, with enormous amounts of energy being liberated in a shock wave that propagates through the outer envelope to produce a gigantic supernova explosion of partially enriched material. Some of the ejecta are in layers that underwent nuclear fusion to form silicon or oxygen, other innermost ejecta are exposed to a high neutron flux and form highly enriched isotopes.

When the core of the star is so highly compressed that electrons are squeezed on to the protons to form neutrons, the ultimate state of compression is when the neutrons are virtually touching. In essence, the star is a giant nucleus. The pressure that supports this ball of neutrons arises from the quantum motions of the neutrons. The neutron core is held up by the degeneracy pressure of the neutrons. This sets in at a density that is a billion times larger than that of the white dwarf, or ten billion tonnes per cubic centimetre. A neutron star is a thousand times smaller than a white dwarf, or 10 kilometres in radius.

Elementary particles have a wave-like nature, and electromagnetic radiation equally can be particle-like, especially at very high energy. The more massive the particle, the smaller its wavelength. It is the wavelength of a particle that determines the uncertainty in its position. The Compton wavelength of a neutron is a thousand times smaller (the neutron mass being a thousand times larger) than that of the electron. This is why a neutron star is a thousand times smaller than a white dwarf. Thousands of neutron stars are detected by radio astronomers as pulsars: rapidly spinning magnetized neutron stars that emit radio beams from particles that are accelerated near their magnetic poles.

The supernova luminosity is comparable to that of an entire galaxy for some months after the explosion. Supernova remnants are responsible for polluting the interstellar gas and recycling the enriched ejecta into successive generations of stars. A dying star once explosively ejected the carbon and iron in our bodies.

The Milky Way: the past unveiled

Galaxies are assemblies of stars and interstellar matter that have incorporated successive generations of supernova ejecta into the heavy element abundances that are measured at present. The stellar record incorporates the entire history of chemical evolution, sampling stars that formed when the galaxy was young and metal-poor, as well as younger, more enriched stars. The abundances of elements in the interstellar medium give a current epoch snapshot of galactic chemistry. This works as follows. Stars form in dense clouds of atomic and molecular hydrogen. The forming galaxy was a collection of many such clouds. Supernovae exploded and enriched each cloud. A few supernovae in any cloud were sufficient to provide enough energy to disrupt the cloud. The supernova ejecta mix into the interstellar gas and are dispersed into the interstellar medium. The early gas-rich disc is gravitationally unstable to the formation of new clouds. The disc breaks up into agglomerations of clouds. New clouds form out of the enriched interstellar gas. These in turn form stars. The process repeats over several billions of years.

We arrive today at a mixture of stars and gas. There are old stars that formed in the first generation of clouds. Two distinctive features characterize these stars. The oldest stars are metal-poor, reflecting the paucity of elements heavier than helium in the early clouds. Moreover, the earliest clouds had not yet formed the galactic disc. They have plunging orbits that carried them in from the outer halo of the galaxy towards the centre, following the overall collapse of the early galaxy. The first stars retain orbits that match the orbits of the clouds from which they formed. As the disc forms, later generations of clouds form on predominantly circular orbits, and settle into a thin disc. The stars that form from the clouds reflect the cloud orbits, so that one expects to find a general correlation between increasing metallicity and lower random motions of the stars out of the galactic plane. Such 'vertical' motions mostly reflect the early collapse and formation of the disc. The youngest stars have motions predominantly in the galactic plane.

Today, the interstellar medium has heavy element abundances that are slightly in excess of solar abundance. Since the Sun formed 4.6 billion years ago, this represents the subsequent enrichment of interstellar gas by supernovae. Approximately 30 per cent of the Milky Way disc is gaseous, the rest being stars. The gas supply would have been exhausted long ago were it not for the ejection of mass as stars evolve and die.

Another window on to the past of the Milky Way has emerged from studying the orbits of stars and globular clusters in the galaxy halo. Aggregations of stars as well as globular clusters are found that are counter-rotating, meaning opposite to the prevailing direction of rotation. These are relics of ancient encounters between dwarf galaxies and the Milky Way. These resulted in the almost complete destruction of dwarf satellites as their gas and stars were incorporated into the Milky Way. The gas is tidally stripped and rains down on to the disc. The dense core of stars spirals into the central bulge as it undergoes a form of dynamical friction with the ambient disc stars. The loosely bound outer stars of the dwarf are also tidally stripped but, unlike the gas, which loses energy by dissipation and falls into the disc, retain their orbital kinetic energy. This is the origin of the counter-rotating stellar streams that are found in the old star populations of the Milky Way halo.

In summary, we could say that our cherished beliefs, not to be abandoned at any price, endorse the Big Bang model back to at least one second. We cannot attribute any comparable degree of confidence to descriptions of earlier epochs because any fossils are highly elusive. Bearing this restriction in mind, we can now assess the paradigms for structure formation. The basic framework is provided by the hypothesis that the universe is dominated by cold dark matter, seeded by fluctuations that were produced at the beginning of the universe. A phase of inflation occurred.

Fluctuations, initially the imprint of quantum fluctuations on infinitesimal scales, were imprinted on to the macroscopic scales relevant for galaxy formation eventually to occur. This happened at a much later stage in the universe, when the matter content was first dominant over radiation. Only then, some tens of thousands of years later, could fluctuations become stronger under the influence of gravity. This scheme does remarkably well at accounting for many characteristics of the large-scale structure of the universe.

Dark matter dominates large-scale structure and accounts for at least 90 per cent of the mass of the universe. The evidence is very strong, and comes primarily from the clustering of galaxies. On smaller scales, the rotation of galaxies has yielded the most robust evidence for dark matter.

Spiral galaxies are elegant examples of rotating systems. The spiral patterns represent compression waves in the interstellar gas that are typically caused by the passage of a nearby orbiting satellite companion galaxy. In the case of the Milky Way, the Large Magellanic Cloud fills this role. As the galaxy rotates, the pattern of compression takes on a spiral form. Stars form in the compressed gas. Luminous associations of hot, newly formed stars demarcate the trailing spiral pattern that is generated by the differential rotation of the galaxy. Differential rotation means that the inner stars lap the outer stars. Typically the rotation velocity is constant: the inner stars have smaller orbit trajectories than the outer stars. Gas clouds are compressed by the density wave that is often induced by close passage of a nearby companion galaxy. The compression of the gas triggers cloud coalescence, which in turn destabilizes the clouds. The result is that stars form.

Dynamical measurements provide a robust approach to measuring dark matter on larger scales. These methods arise from 'weighing' large regions of the universe, notably galaxy clusters and superclusters. Determination of the random motions of galaxies enables one to estimate the local mass that is responsible for the overdensity in the universe that drives these motions. Without such an overdensity, one would have a perfectly uniform expansion. Each galaxy would have strictly the recession velocity, no more, no less. In practice, the density irregularities and the observed galaxy structures induce slight deviations from the uniform expansion. Studies of galaxy clusters measure the mass on scales of about 1 megaparsec, but one can use the random components of galaxy motions to measure the mass density on scales of up to 30 megaparsecs. One infers that the mean density of the universe is about 30 per cent of the critical density. Most of the matter in the universe is not baryonic.

Note

1. See John Mather and John Boslough (1996) *The Very First Light*. New York: Basic Books.

14 Beyond the Beginning

In my end is my beginning.

Mary Stuart

What we call the beginning is often the end
And to make an end is to make a beginning.
The end is where we start from.

T. S. Eliot

Quantum uncertainty is the key to understanding what, if anything, happened before the beginning; that is, time zero as measured by the cosmologist. The logic is worth examining. The universe is expanding, and therefore it had a beginning. Extrapolate known physics towards the beginning, and we reach a point of failure. Classical physics is inadequate to cope with the extremes of density and temperature that we know must have been present. So far, so good. Enter quantum cosmology: here is where the fun begins.

Beyond creation

We have sought for firm ground and found none. The deeper we penetrate, the more restless becomes the universe; all is rushing about and vibrating in a wild dance.

Max Born

Quantum uncertainty is the essence of what quantum theory has to offer cosmology. The uncertainty principle has been well tested on subatomic scales. It explains why particles are sometimes particles and sometimes waves, manifest in such devices as the electron microscope. A wave can weave its way around atoms, through seemingly solid matter, a phenomenon known as quantum tunnelling. Imagine the chair on which you are sitting tunnelling through to the floor below. Quantum theory says that this could happen, though the probability of such an occurrence in your lifetime, or even that of the Earth, is infinitesimal. Yet this seeming reluctance of quantum fuzziness to exert itself on macroscopic scales has not deterred cosmologists from postulating that cosmic fuzziness applies on cosmological scales, and is the supreme arbiter of the beginning of time.

Here is how it all began. The eminent relativity theorists John Wheeler (to whom we are indebted for coining the phrases 'black hole' and 'wormhole') and Bryce DeWitt were at a loose end on some, one can only speculate, taxing social occasion in the 1960s when Wheeler posed the problem: 'We demand of physics some understanding of existence itself.' Their solution: take the equation for a wave function of an atom, which expresses the atom's fuzziness in position and time, and reformulate it for the entire universe, envisioned as some sort of superparticle. It took only another decade or two before James Hartle and Stephen Hawking solved the Wheeler–DeWitt equation to derive what they boldly asserted to be a prescription for the wave function of the universe. Armed with this, they could predict the present state of the universe, and in so doing evade all the issues surrounding the moment of creation and the initial singularity. Fuzziness was elevated into a supreme role, which was just as well given the scarcity of alternative hypotheses.

Of course, once the quantum Pandora's box was opened, there was no going back. An electron cannot simultaneously be particle and wave. Who chooses? One widely accepted interpretation blames the observer. Until he or she looks, a quantum system has neither state. One interpretation of quantum theory is that the act of observation resolves the system into a particular state. Until the observer performs the act of observation, no prediction is possible. This is true at the quantum level, because of the uncertainty principle. One can never precisely locate a particle. But quantum phenomena can affect macroscopic objects, bizarre though this seems.

The best known example is that of Schrödinger's cat, a thought experiment devised by the German physicist Erwin Schrödinger in the 1930s. The idea was that the radioactive decay of an atom triggered a Geiger counter, which in turn was connected to a machine that, when triggered in turn by the Geiger counter, released poison gas into a cage in which the cat was happily purring. Now suppose that the radioactive isotope has a half-life such that after one minute, there is a 50 per cent chance of a radioactive decay occurring. This is intrinsically a question of probability according to the quantum theory, which cannot provide any more precise answer. This is rather hard on the cat, however. The cat has a 50 per cent chance of being alive and a 50 per cent chance of being dead. This of course makes little sense: a cat is either alive or dead.

Quantum theory connects the life-or-death paradox for the cat with the microscopic uncertainty that threatens its life, and indeed its sanity, by postulating that no one will ever know if the cat is alive or dead until the cage is opened and the cat is observed. At this point there can be no doubt: the cat is either dead or alive, and both possibilities carry equal likelihood. The Danish physicist Neils Bohr argued that the act of observation crystallizes the state of the cat into one of life or death. Only then is macroscopic reality meaningful. Before the act of observation was performed, the cat had the potential of being alive or dead, but its state was indeterminate.

Many physicists are unhappy with this interpretation of quantum reality being dependent on the act of observation at the macroscopic level. The question was asked: what state was the cat in before it was observed? Presumably either alive or dead, or did it simply not exist? The idea that only what we observe actually exists is beloved by some philosophers, but does not satisfy physicists, who prefer to believe in the objective nature of reality. The dilemma is unresolved if quantum acts have macroscopic consequences.

There is a way out, perhaps, which is simply to argue that the quantum world disconnects from the macroscopic world. Any quantum effects are immeasurably small. Quantum probabilities vanish, to be replaced by reality. However, this pragmatic approach has to cope with a remarkable series of experiments, which purport to show that quantum effects can act over kilometre scales. The story begins with a thought experiment due to Einstein and his collaborators Boris Podolsky and Nathan Rosen. Einstein was a great sceptic about the quantum theory and set out to develop a paradox that was seemingly so absurd as to refute it. Radioactive decays produce spinning particles, but do not generate any net spin – the spin of the particles cancels out. Consider, for example, a pair of particles ejected by a nucleus, consisting of an electron and a positron. One particle spins one way, the other in the opposite direction. To conserve momentum, the two particles move off from the parent nucleus in opposite directions. There is no net spin or momentum.

Now imagine two detectors designed to capture the electron and the positron, and to measure the spin. The first detector is switched on. A particle arrives. It is a positron with, say, positive spin. Then the remote detector, which is turned on slightly later, measures the accompanying electron . . . but the spin is known before the measurement is done. It must be negative. So quantum laws, which insist that the first measurement could not be predictable, nevertheless allow a prediction of the second measurement performed kilometres away. Before the first measurement was done, all one could say was that the second detector had a 50 per cent chance of measuring a certain outcome, yet once the first detector did its measurement, one could be 100 per cent confident of the outcome of the measurement by the second detector. Actual experiments have been performed that demonstrate this effect over distances of hundreds of metres or even of kilometres. Does this tell us that quantum phenomena have macroscopic consequences? Could Schrödinger's cat survive in some new state? The jury is still out on this question.

This hardly helps us with the very early universe, when observers were exceedingly rare, and neither observation nor measurement could have been fundamental elements of the theory. The resolution was not observation, but lay in the concept of observability. This may seem a subtle distinction, but the consequences are shattering. Now one has an infinity of possibilities, and an infinity of histories. What subset of this mind-boggling vista of events actually occurred is a question that is answerable, at least in principle, and one that can be posed to astronomers. Of course, whether they can provide any meaningful answers is debatable.

Parallel universes

> I can safely say that nobody understands quantum physics.
>
> Richard Feynman

The existence of many options can be taken to mean that there are multiple universes, all part of some higher dimensional superspace.

One striking implementation of the concept of multiple universes begins with complete chaos. Any small region in the primordial quantum chaos has the potential to breed a new universe. A tiny patch of space inflates. Some patches inflate a little, some a lot. One particular patch is very successful in inflating, and becomes the dominant universe, ours.

The origin of the universe is explained, and not assumed, by the postulate that quantum chaos breeds inflating bubbles, within which further bubbles develop and inflate. Chaotic inflation provides the gateway to creation of multiple universes. Random parts of the early universe may inflate by random amounts. Universes are born in profusion. The universe may exist in an infinity of incarnations, each one subtly different. Quantum cosmology, in at least one of its variants, asserts this to be the case. Any one of these could be our universe, except that our observed universe is remarkably vast compared to the scale of things near the quantum era. We are the product of a spontaneous fluctuation in an eternally self-reproducing universe. All it takes is one event, however unlikely, to spawn our universe. So we require an exceptional bubble, but there is an infinite time available to await its spontaneous birth. There is an infinity of universes, all but one of which are forever inaccessible to us. The most probable one is the largest, and by a sort of Darwinian survival of the fittest, this is the universe that outgrew its rivals to end up as the most likely candidate for our very own universe.

Another perspective on multiple universes arises in the many-worlds hypothesis of the American physicist Hugh Everett, who asserts that quantum phenomena continuously give birth to parallel universes. These universes are more than magnificent hypotheses. This interpretation has mind-boggling consequences. Every quantum option spawns a universe. There are untold billions of parallel universes, all equally real and viable, but with no possibility of any physical communication. Multiple universes are all equally real until an observation takes place. They occurred and exist, at least until the observer points the telescope, when just our own boring patch of spacetime is seen.

The role of consciousness

Others would prefer to argue that such multiple universes are not necessarily real. One quantum view asserts that reality requires an observer. Even this seems rather odd to some physicists. Such a viewpoint brings in consciousness as part of the definition of reality. There are physicists, such as Roger Penrose, who have argued that some as yet unknown physics may be required in order to understand consciousness. As yet there is certainly no indication that consciousness even lies within the realm of physics. Nor, to be fair, is their any indication that it does not. However, my own view is that enlightenment will come, some day, via the science of biophysics. No mystical effects will be needed. Consciousness is acquired just as a newborn baby or even embryo develops a functioning brain. The closest attempts to combine consciousness and quantum physics are found in the almost mystical concept of hidden variables, due to quantum physicist David Bohm, that act as the observer's ghostly all-pervading hands, which instantaneously pluck the measured state from the myriad of alternative forms. Here, if not before, is where metaphysics and science begin to overlap. Once the scientists dabble in mysticism, there is no stopping the philosophers, the theologians, and the cohorts of amateurs, with their own pet theories, who are clamouring to breach the gate to cosmology. Indeed, the vast majority, as exemplified by the success of astrology columns in the press, would question the credentials of the tiny minority of scientists to make any such value judgements. Who are we scientists to say whether quantum weirdness has anything to do with reality? I beg to differ. It seems to me that a solid training in science, and especially in physics, does give one enough perspective to discern reality from myth, albeit dimly.

The multiverse

Reality is something that astronomers can hope to measure and probe. There is a connection between multiple universes and observation, although farfetched. British astronomer Sir Martin Rees defines the multiverse to be the ensemble of all possible universes. Let us then ask: can

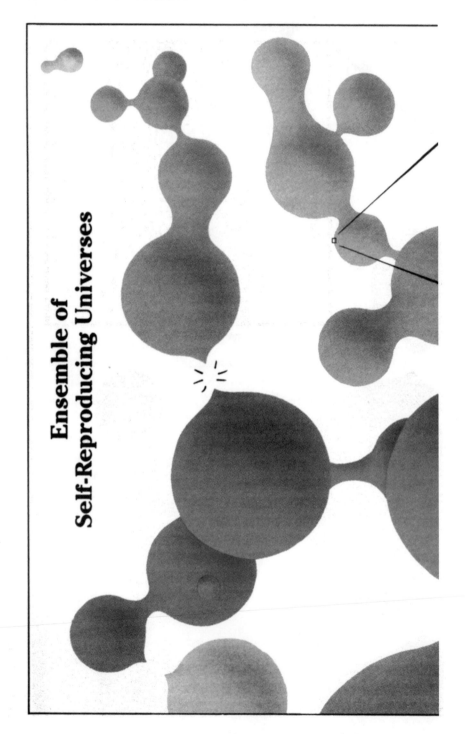

Ensemble of
Self-Reproducing Universes

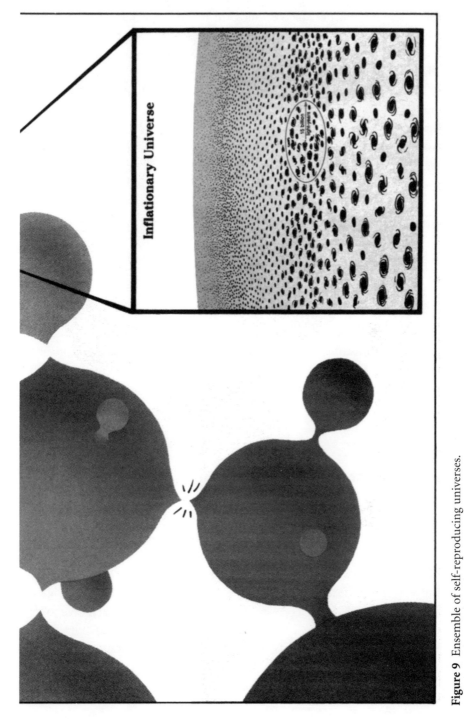

Figure 9 Ensemble of self-reproducing universes.

Source: E. Mallove (1988) 'The self-reproducing universe'. *Sky and Telescope*, September.

these constitute a part of cosmological reality? It would seem that such a metaphysical agglomeration of universes is inherently unobservable, because all are inimical to intelligent life but for a very small subset, in which is featured our own delicately balanced cosmos.

Nevertheless, the many-universe postulate is intellectually challenging, and purports to explain a plethora of unlikely circumstances. Why is the essence of life, the carbon atom, poised to be formed by capture of its constituent atoms, which otherwise would fly apart in the intense fusion reactions under way inside a stellar core? Why is the vacuum empty, or almost so? Why are the fundamental forces so distinct in strength from each other? In particular, why is the strong nuclear force neither weaker nor stronger, when, if it were, stars would not form? Why is the weak nuclear force as weak as it is, when otherwise elements would not form? Why is the neutron just 14 per cent more massive than the proton, which it needs to be for the formation of hydrogen? Why is the electron mass only 1/1836 of the proton mass, when if it were much larger, molecules such as DNA would not form? Why are fluctuations small enough, but not too small, to allow the formation of galaxies? The list seems endless. How easy it is to postulate an infinity of universes that violate these precepts, and thereby remain unobserved, devoid both of cosmologists and their books! Surely there must be some underlying theory, which provides a physical explanation.

The birth of baby universes

The American physicist Lee Smolin[1] posits that black holes provide the underlying solution to a physicist's dream of the final theory. Embedded within black holes may be other universes. A wormhole, which connects a black hole to its opposite, a white hole, is the port of entry to a new universe, newly formed, or perhaps ancient, and remote, or possibly not, in space and time from its parent universe. Every time a black hole forms, a new universe is spun off deep within. There are untold numbers of black holes in the universe. Perhaps there is an infinity of universes within universes.

Within these universes, the constants of nature suffer small but random changes every time a black hole forms and a new universe develops. Baby universes undergo evolution. Spin-off universes could display the effects of heredity and environment. The most probable universe emerges from this evolutionary sequence with the maximal number of evolutionary pathways, and hence black holes. We are at the peak of a grand cosmic summit of natural selection of the laws of physics. And, here is the beauty of the argument, such a course of events is inevitable: we are the outcome. This is not a bad postulate if it is testable, which Smolin argues to be the case. For example, one needs only to search for the black holes. Optimal procreation would require a furious proliferation of black holes, a conceivably testable prediction by X-ray astronomers.

It is always wise to inject a note of harsh reality before constructing too elaborate an edifice. We do not yet understand how stars form in nearby nebulae, although we are getting closer to this goal as observational techniques and telescopes improve. Astronomers believe that every star above about 25 solar masses is destined to end its life as a black hole. But to prove that this is inevitable is quite another matter, as also is the observability of the resulting black holes. Maybe you see them, maybe you don't, depending on whether or not they are accreting gas at a great rate. Some may form via fiery explosions, some may form quiescently. Many, we believe, are visible because they are bathed in an envelope of glowing, X-ray emitting gas spewed out by an accompanying low mass star in a close orbit.

But it may be that the situation in a star-forming cloud is so complex that it will take the power of the supercomputers of the far future to be able seriously to address the physics of what determines the mass of a star like the Sun. Even weather prediction over a period longer than a week or so is a gamble. If we cannot model meteorology well enough to predict long-range weather patterns, one has to retain a sceptical attitude to the alleged need for a multitude of other universes that derives from poorly understood phenomena in our own universe.

Note

1. See Lee Smolin (1997) *The Life of the Cosmos*. Oxford: Oxford University Press.

15 Towards the Infinite Universe

Space is almost infinite. As a matter of fact, we think it is infinite.

Dan Quayle

... the existing scientific concepts cover always only a very limited part of reality, and the other part that has not yet been understood is infinite.

Werner Heisenberg

If the universe is flat, it is infinite. That at least is what the textbooks tell us. The concept of an infinite universe, or at least of one that is virtually infinite, has received support from the cosmology of inflation. Our universe underwent a phase of exponential growth. There is every reason to expect that it underwent an expansion in scale by many powers of ten within a fraction of a second. Prior to inflation, all the matter within the visible universe spanned less than the size of a pea. Even if the entire universe had begun as a finite space, its present dimension is likely to be at least ten to the power of one hundred times larger than our visible horizon.

What is infinity?

It is not easy to define infinity. Some philosophers endorsed a theological connection. 'Every relationship of man to the infinite is religion', according to the German philosopher Friedrich von Schlegel (1772–1829), who, however, made an exception for mathematicians – 'Whenever a mathematician calculates infinity, that, to be sure, is not religion' – and certain of his contemporaries: 'there are books in which even the dogs make references to the Infinite'.

Many, from poets to pundits, and especially theologians, are uncomfortable with the concept of an infinite universe. Bishop Ernest Barnes (1874–1953) was a mathematician who held high office in the Church of England. In 1931, he participated in a public debate with cosmologists, and commented that 'It is fairly certain that our space is finite, though unbounded. Infinite space is simply a scandal to human thought . . . the alternatives are incredible.' Perhaps what he had in mind was that in an infinite universe there could be another Bishop Barnes holding opposing views. But if such a person existed, he would have to be almost infinitely far away. Hence his existence is completely irrelevant, and the heavenly twins could never have been aware of one another's existence, let alone communicated.

Dreams of infinity occasionally ended badly, as happened to the Italian philosopher Giordano Bruno (b. 1548), who was burnt at the stake in Rome in 1600 for his challenge to papal authority, which focused on his insistence that 'There is a single general space . . . in it, are innumerable globes like this on which we live and grow; this space we declare to be infinite.' Poets, especially of the Gallic variety, seem equally appalled. 'Malgrè moi l'infini me tourmente' ('I can't help it, the idea of the infinite torments me'), wrote Alfred de Musset (1810–57). And two centuries earlier, the scientist and philosopher Blaise Pascal (1623–62) said 'Le silence éternel de ces espaces infinis m'effraie' ('The eternal silences of these infinite spaces terrifies me').

There are certainly theological issues to be raised about the infinite. The probability of intelligent life, or grand design, may be infinitesimally small, but it inevitably would have occurred somewhere at some time in an infinite universe. And here we are. The rest of the universe is irrelevant, since there is no one around to observe it. This is an anthropic argument, but one so weak that it is almost a tautology.

In contrast, a cheery ray of optimism emanates from the green pastures of England. William Blake (1757–1827) is able

> To see a world in a grain of sand
> and a heaven in a wild flower,
> hold infinity in the palm of your hand
> and eternity in an hour.

Needless to say, the poets claim precedence over any theological views of the infinite. For example, the American-Russian poet Joseph Brodsky (1940–96) wrote, 'The poetic notion of infinity is far greater than that which is sponsored by any creed.'

Scientists seem to be more at ease with an infinite universe than poets and theologians. Isaac Newton (1643–1727) was perhaps the first to address the notion of gravity in an infinite universe, when he wrote in a letter to Richard Bentley in 1693:

But if the matter were evenly disposed throughout an infinite space . . . some of it would convene into one mass and some into another, so as to make an infinite number of great masses, scattered great distances from one another throughout all that infinite space. And thus might the sun and fixed stars be formed.

As a result of the successes of Newtonian gravity, Newton's cosmological views were broadly acccepted: consider Immanuel Kant (1724–1804): 'In the infinite distance, we see the first members of an unbroken series of stars. . . . There is no end here, there is a truly immeasurable chasm. . . . The universe is filled with worlds without number and end.' Not surprisingly, the Greeks were there long before: 'There are worlds infinite in number and different in size', according to Democritus of Abdera (460–370 BC). Epicurus (341–270 BC) concluded that

the universe as a whole is infinite, for whatever is limited has an outermost edge to limit it, and such an edge is defined by something beyond. Since the universe has no edge, it has no limit; and since it lacks a limit, it is infinite and unbounded. Moreover, the universe is infinite both in the number of its atoms and in the extent of its void.

This theme was repeated by Lucretius (96–55 BC), the Roman poet and follower of Epicurus:

> The universe is not bounded in any direction.
> If it were it would necessarily have a limit
> somewhere. But clearly a thing cannot have a limit
> unless there is something outside to limit it.
> In all dimensions alike, on this side or that,
> upward or downward through the universe there is no end.

Nor were the Greeks even first to explore infinite worlds. According to the Egyptian creation myth, the universe began 'as a primeval sea unlike any sea which has a surface, for here there was neither up nor down, no distinction of side, only a limitless deep – endless, dark and infinite'.[1]

Modern cosmology echoes many of these ideas on infinity. There is no proof that inflation ever occurred, or, indeed, if it did occur, that its influence extended well beyond our observable horizon. An important myth to overcome is the inference that a flat or even a negatively curved universe is infinite. This issue of size must be the core of cosmological tests. How can we make a hard scientific appraisal of whether the universe is infinite? It turns out that the question of whether or not the universe is nearly infinite can be experimentally tested. The universe can be very large indeed, compared to the visible scale, yet its size can, in principle, be measurable.

Before the Big Bang

> What did God do before He made heaven and earth? He was preparing hell . . . for pryers into mysteries.
>
> Saint Augustine

The term Big Bang suggests that the universe began with an explosion. But cosmologists often reject the concept of an explosion, and even the word itself. Cosmologists do not like the term explosion because it conveys the idea of sound, like a 'bang!', and it makes no sense to think of that. But apart from this quibble, the word explosion is valid. The simple description of how the universe originated is an explosion, in the sense that it began from a very small volume and increased very rapidly. The Big Bang starts with an 'inflation', a short period during which the universe expanded enormously at very high speed. But what happened before that period? Perhaps long before inflation there was a universe that was collapsing near a singularity, and then inflated again, so there was already a history before the Big Bang. Some people think there was a 'pre-Big Bang'.

The problem is that the universe was thoroughly devoid of any order at the beginning. This means that the initial entropy of the universe was very large. Yet we as cosmologists like to think that order developed with time, as galaxies formed. How does one square this late-time entropy content with the enormous primordial entropy? The answer may lie in the formation of black holes. These can swallow entropy, along with everything else, and so can generate a universe resembling our own. Galaxies form, but so do black holes. Perhaps on balance not much changes. One has to hope that when the black holes finally decay, they do not regenerate the entropy they consumed. Or by then, it will be far too late to make any difference. Even protons will have decayed.

What happened before the Big Bang is a subject of intense speculation. There are conceivable traces. One is an imprint on the fluctuations in the cosmic microwave background radiation. This is almost as if someone wrote in the sky: 'Look, I was here!' The point is that in principle density fluctuations might have survived the Big Crunch, if there was one. Exactly how they are modified is unknown. There is no physical way to explain the transition from pre- to post-Big Bang. We have no understanding of how to change from collapsing to expanding. There are predictions of what the microwave background might resemble were there a pre-Big Bang. Needless to say, no evidence has yet been found that favours a pre-Big Bang. Despite all of this, we return below to the possibility of a cyclic universe.

A cyclic universe

The idea that the universe may have an infinite age and that the present phase is one cycle of an infinite number is an idea that has its roots in ancient mythology. Big Bang cosmology ran into several major obstacles when it tried to incorporate the analogous idea of a series of big bangs in alternation with big crunches. Each Big Bang would breed a host of stars and galaxies that in the succeeding Big Crunch would get compressed and pulverized into radiation, all, that is, except for the many black holes that would also have formed.

The entropy, or randomness, of the universe would increase from cycle to cycle, unless there was some way of unlocking all the information lost by collapse into black holes. This seems very unlikely with known physics. The increase in entropy means that the heat content of the universe increases from cycle to cycle. The Big Bang would therefore achieve ever-larger radii before recollapsing in successive cycles. Only a finite number of past Big Bangs could have occurred. And there would be many relic black holes from previous cycles. The universe would be a strange place, crammed with black holes, and could not have been in existence forever.

For the Big Bang to reverse itself, the mean density of matter in the universe must exceed the critical density. We know that the density is only one-third of the critical density. And that includes all the dark matter and black holes. So a cyclical universe would seem destined for the dustbin. However, little is definitive in cosmology, and old ideas have a habit of reasserting themselves. Two developments have revived interest in a universe that can perpetually reinvent itself. One is the notion of decaying dark energy. Dark energy, if it is constant, is just the cosmological constant term that Einstein invented to stave off collapse of the universe. Observations of very distant supernovae have shown that the universe is presently accelerating, and dark energy is precisely what is needed to accomplish this. There is no theory for dark energy, so cosmologists are happy to develop forms of dark energy that are even more exotic than Einstein's cosmological constant term.

The motivation is the following. In the presence of Einstein's cosmological constant, the universe has recently entered a phase of acceleration. It will continue to accelerate and expand forever. The future will be very bleak and empty. A chilling prospect awaits us.

Enter a concept from the theory of quantum gravity. According to this, dark energy is a manifestation of higher dimensional gravity. It fades away with time. And the resulting universe is thoroughly empty. All the black holes are exponentially far away. But this emptiness is dangerous, as it has the potentiality of triggering a collapse phase. The universe goes from Big Bang, when it is indefinitely large, to Big Crunch, when it recollapses to a near singularity. And at this point, the re-expansion is initiated. There are an infinite number of cycles of an infinite universe.

No one has the remotest idea of why this works. But one could have said much the same about inflation. Suffice it to say that time goes on relentlessly. Physics does not allow us to reset the clock. The cyclic universe is infinitely old. And it is infinitely large. It cleanses itself while expanding exponentially, and regenerates entropy while collapsing. The cleansing process works by dilution. Exponential expansion means that the dark energy dominates over the energy density in radiation, in matter, and in such objects as black holes that would be relics of star and galaxy formation. Ordinarily, the inevitable consequence of the second law of thermodynamics results in the increase of entropy as stars evolve and die. This has been said to lead to the heat death of the universe. All is changed in a cyclic universe, however. Eventually, by the time that collapse occurs, the entropy clock is reset to zero.

A cyclic universe is an attractive idea. But there is little incentive to prefer it over one that underwent a single Big Crunch from an infinitely large state, before bouncing to produce the Big Bang. The pre-Big Bang phase may leave relics for eventual discovery, such as black holes pro-duced in the collapse, or it may not. And then again, who is to say that the universe is infinite? This is unprovable, and it may simply be very large.

Ghosts

What determines the size of the universe is a geometrical property that is known as its topology. Topology describes the degree of connectivity of space. A coffee cup differs from a wine glass. A wine glass topologically is a sphere; a coffee cup is a toroid. And a cylinder differs from a toroid, although both are spatially flat (in the sense that Euclidean geometry is the norm). It is a curious fact that the theory of gravity says nothing about the topology of the universe. Einstein's brilliant identification of gravitation with geometry produces a theory of space and time that is global in its conception. All of space–time is included, inextricably interwoven with gravity.

> Space-time grips mass, telling it how to move.
> Mass grips space-time, telling it how to curve

as so eloquently phrased by John Wheeler.[2] But the global topology of the universe remains unspecified. The universe need not be infinite, but merely very large. The universe could be flat, but have the topology of a giant toroid or doughnut. The surface of a toroid is geometrically flat, in the sense that parallel lines stay parallel. Or the universe could resemble the topology of a Klein bottle, a sort of twisted toroid. Of course, the surface of the toroid represents a two-dimensional analogy to the three-dimensional space of the universe. It is the overall pattern that matters. Topologists are somewhat limited. A topologist cannot distinguish a doughnut from a coffee cup.

Given the richness of possible topologies of the universe, it seems reasonable to infer that a flat universe can be, and even plausibly is, finite. It may, of course, be very large indeed. It turns out that the question of whether or not the universe is nearly infinite, or more precisely how large it is, can be experimentally tested. The universe can be very large indeed, compared to the visible scale of the horizon, yet its size can, in principle, be measurable. If the universe really were infinite, this would not easily be testable, but we could at least set a lower bound on the true size of the universe.

Suppose the observed universe were represented in two dimensions by a small patch on a giant torus. This would be a compact topology, as opposed to one that was infinite in one direction, a cylinder, or in two directions, a sheet. Of course, three-dimensional spaces have more variety, but the principles are the same. If the topology of the universe were compact, light could propagate in circles. This might take a long time, depending on how large is the circumference of the torus on the surface of which the photons pass. We have no a priori knowledge of how large this topological scale is. If the scale of the topology were smaller than the horizon of the universe, there would be enough time for light to have propagated away and returned by a different path. One could in principle see the back of one's head! More realistically, we would be able to see ghost images of galaxies.

One advantage of a small universe is that because light has had time to propagate many times around the universe at any given epoch, any early irregularities or anisotropies would have been erased. This provides a potent alternative to inflation for erasing all memory of initial conditions.

The topology of a flat universe may be even more complex. Imagine a multiholed torus, in effect a pretzel rather than a doughnut. In this case, the sky would contain galaxy ghosts galore, multiple copies of the same image. It might be difficult to disentangle this information. After all, one galaxy looks like another, especially after the inevitable blurring caused by gravitational lensing of the light rays by intervening galaxies, and the scattering of the light by intergalactic dust. There is another hope, however. Imprinted on the fossil fluctuations in the cosmic microwave background radiation is a pattern due to the topology of the universe. Every topology has a unique pattern. Nor does the scale of the topology need to be smaller than the visible horizon. Even large topologies leave their mark.

Seeking the topology of the universe

The universe may be a gigantic torus, or at least the three-dimensional generalization of a doughnut, a hypertorus. A hypertorus universe has bizarre implications. Such a universe is spatially flat, locally Euclidean. The universe must contain standard inflationary fluctuations, required to generate large-scale structure. Normally, these fluctuations are described by a random density field, but a novel feature is introduced by the compactness. Even with random fluctuations, there are preferred directions because of the torus-like topology. Some fluctuations can be extended along the long circumference of the torus, and others are limited by the short circumference. An observer of the cosmic microwave background radiation will see locally random fluctuations in a universe that will look like the usual Big Bang of Friedmann and Lemaître. However, globally the universe will seem to be anisotropic.

This is because light can propagate on the surface of the hypertorus to get from one point back to itself either in a short circle, perpendicular to the spline of the torus, or along a long circle, parallel to the spline of the torus. The different path lengths for light lead to patterns imprinted on the cosmic microwave background sky because of the anisotropy. Different photon paths from a given scattering event have differing lengths, so they lead to images of the same source but with different intensities and in different directions.

If there is time for the photons to propagate around the universe, this leads to ghost images. Now imagine averaging over an ensemble of observers. The net effect will be that the cosmic microwave background sky will seemingly have non-random temperature fluctuations. There will be patterns on the sky. In this way, one can hope to observe the topology of the universe via its patterns on the sky. Of course, non-random cosmic microwave background fluctuations are something of a minor industry, since they can be produced by a host of other mechanisms. The artefacts arising from looking at the microwave sky through our Milky Way galaxy induce temperature variations. These reflect the non-random geometry of the interstellar medium, such as that of filaments and sheets of diffuse emitting gas or plasma. Complex scenarios for inflation can also generate non-random patterns on the sky.

How does one sort any of this out? Light from any source spreads out into a spherical wavefront. Imagine some randomly placed observer at the epoch of the last scattering of the radiation. Her horizon can be depicted as a circle on the sky, a sphere in three dimensions projected on to the celestial sphere. In a large universe, this circle spans the entire universe. But in a topologically small universe, one would expect to see many circles on the sky, in the form of a circular pattern of infinitesimal temperature variations.

These circles on the sky are not seen: the universe cannot be very small. If the topology scale is less than the horizon scale, there will be an apparent truncation in the distribution of scales associated with temperature fluctuations. At some level, nothing like this is seen, although there is an unexplained deficit of power on the largest scale (that of the quadrupole). The universe must be really rather large. But how large?

Flatness of space makes life simpler for the seeker of cosmic topology. If the geometry of the universe were hyperbolic, there would be infinitely

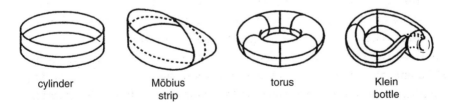

| cylinder | Möbius strip | torus | Klein bottle |

Figure 10 There are several different possibilities for the topology of a flat universe. In these two-dimensional examples, space is represented by multiply connected structures whose surfaces are flat but have domains that are inaccessible. In a simply connected topology, such as a sphere, every point is accessible.

many topologies. The topological choices simplify in a flat universe. There happen to be just 18 distinct types of flat spaces. All of these, except for the analogue of the infinite plane, are multiply connected; that is to say, closed loops are possible. Only ten of these spaces are compact, the others being infinite in one or more directions.

We can do even better in terms of limiting ourselves to physically meaningful compact space. It seems reasonable to demand that space–time be 'time-orientable'; that is, there is an arrow of time and an orientation of space. There should be a past and there should be a future. Left should stay left, and right should stay right as we voyage in time, even hypothetically, around the universe, returning to our point of departure. If time is orientable, so must be physical space. In other words, go around the universe, and when you return you have preserved your handedness or parity, unlike your reflection in a mirror. Our hypothetical space voyager should not have to cope with meeting her mirror image. Fundamental concerns about physics leave us with the requirement that the combination of space and time, applicable to our space–time voyager, must be totally orientable.

Of the 18 possible flat spaces, only the ten compact spaces avoid infinity in any direction. The compact spaces do have the advantage that one can do certain calculations without running into mathematical infinities. The universe has a finite boundary, which means that the amount of information specified at any time is finite. It makes the future less exciting, perhaps, in the sense that it may be predictable. It is impossible to say whether this is a desirable attribute of the universe without venturing into metaphysical territory. What is potentially intriguing is that the finiteness of the universe may be verifiable by experiment.

The compact spaces include the generalizations to three dimensions of surfaces such as that of the sphere, the toroid, or the Klein bottle. Of these, only six allow one to preserve handedness; that is, left stays left and right stays right on any closed curve. These are the only spaces we can live with if we are to avoid unacceptable paradoxes. This leaves just six physically allowed compact, orientable Euclidean topologies to explore. All the others are unphysical. Each of the possible spaces leaves a distinct pattern on the cosmic microwave background. We can imagine doing an astronomical test by seeking such a pattern. One day, the topology of the universe may be measurable, if it is not too large.

Is the universe finite or infinite?

We do not know if the universe is finite or infinite. The simplest theory of the expansion says that the universe could expand forever, and observations suggest that it should. This corresponds to what is called a 'flat' universe. But the universe could also be finite, because it could be that the universe has a very large but finite volume now, which will increase, so only in the infinite future will it actually be infinite. This is probably just as well, since if the universe were infinite, we could never devise an experiment to prove this.

To give an example, imagine the geometry of the universe in two dimensions as a plane. It is flat, and a plane is normally infinite. But you can take a sheet of paper (an 'infinite' sheet of paper) and you can roll it up and make a cylinder. You can roll the cylinder up again and make a torus (a torus is like a hollow doughnut). The surface of the torus is also spatially flat, but is finite. Actually we had to stretch the cylinder along its length to get it to close into a torus, but the process of stretching does not change the topology.

So there are two possibilities for a flat universe. One is infinite, like a plane, and one is finite, like a torus, which is also flat. Remember that 'flat' is just a two-dimensional analogy. What we mean of course is that the universe is 'Euclidean', such that parallel lines always run parallel, and that the angles of a triangle add up to 180 degrees. Now, the two-dimensional equivalent is a plane, or an infinite sheet of paper. On the surface of the plane you can draw parallel lines that will never meet. A positively curved geometry would require a sphere. If you draw parallel lines on a sphere, these lines will always meet at a certain point, and if you draw a triangle its angles add up to more than 180 degrees. So the surface of the sphere is finite but not flat. The surface of a torus, on the other hand, is a finite and flat space.

Balloon, ground-based, and space experiments have meanwhile extracted much of the data on the microwave sky, which, as we have noted, bears information about the geometry of the universe. The experiments have confirmed that the universe is 'flat'. Even with this knowledge in hand it is not possible to find out whether the universe is finite or infinite.

If the universe is finite, this means that in a two-dimensional geometry it would be like a torus. In such a universe, light travelling on the surface of a torus can take two paths: it can go around the sides of the tube, but it can also go in a straight line. This means that if the universe is geometrically like a torus, light can have different ways to get to the same point. You can have a long way and a short way. This would not be true on a plane. A torus means that space is more complicated. The technical term is that space is multiply connected: there are alternative routes to the same destination. The natural question that now arises is: how small could the universe be, as permitted by the cosmic microwave background observations? Remarkably, even if the topological scale exceeds the horizon scale, there are still weak patterns in the sky.

This is a remarkable implication. It is possible in principle to measure the size of the universe unless it is too large, in which case the predicted patterns are simply too weak. If the universe is like a torus we can see a pattern unique to the finiteness of space. If the universe were finite it could be a hundred times larger than the horizon, which is the distance the light has travelled since the Big Bang, and leave a characteristic pattern. This pattern corresponds to the size of the 'doughnut' or the torus.

The WMAP satellite has measured the cosmic microwave background, the fossil light that filled the universe after the Big Bang, with the highest accuracy ever achieved. Cosmologists have searched for the pattern of topology in the sky. Hitherto, the results have been negative. We conclude from such considerations that the topology scale exceeds about 90 per cent of the present-day horizon scale.

Perhaps the universe will be shown to be measurably compact. This would provide a wonderful triumph for metaphysical and theological reasoning. The ghost of Bishop Barnes would rejoice. But we may be unlucky. If the universe were truly infinite, then we would see no signal at all from topology. All we could really say in this case is that the universe is larger than a certain size. Only if it were finite would it be measurable. And if inflation occurred, the odds are that the topology scale greatly inflated to many horizon scales, and we would never detect it.

Notes

1. J. M. Plumley, cited in L. M. Krauss (1989) *The Fifth Essence: The Search for Dark Matter in the Universe*. New York: Basic Books, p. 5.
2. J. A. Wheeler (1990) *A Journey into Space and Time*. London: W. H. Freeman.

16 From Time to Time Machines

Had we but world enough, and time . . .

Andrew Marvell

We feel that even when all possible scientific questions have been answered, the problems of life remain completely untouched. Of course there are then no questions left, and this itself is the answer.

Ludwig Wittgenstein

Scientists are not very concerned about the philosophical or metaphysical implications of an infinite universe, but are willing to debate the limitless opportunities for the development of alternative forms of life and society. I will borrow freely in the following example of what may seemingly be more science fiction than science, but with ideas that are firmly rooted in physics and biology.

First we should address the key issue that faces would-be time travellers. Time travel leads to the matricide paradox. Suppose one were to meet one's mother as a young girl, and kill her. The logical inconsistency seems a devastating argument that would prevent such a journey from ever occurring. The matricide problem has a possible solution. Consider a billiard ball on a trajectory that allows it to fall down a wormhole. It re-emerges in its past, to collide with its earlier self, knocking itself on to a different trajectory. But this cannot happen. Nature, or rather physics, abhors and, some say, forbids causality violation. The wormhole's gravity defocuses the billiard ball's path. When it emerges, the ball has a high likelihood of missing itself.

In fact the situation is not so different from the everyday problem faced by quantum physicists, of whether Schrödinger's cat is dead or alive. While many physicists have learned to live with this statement, it is of little comfort to a biologist who is accustomed to observing cats that are either alive or dead. Quantum uncertainty may even have resolved the matricide puzzle, at least for the physicist.

Life in an infinite universe

> Given so much time, the 'impossible' becomes possible, the possible probable, and the probable virtually certain. One has only to wait: time itself performs miracles.
>
> George Wald

Other intelligent life forms are inevitable in an infinite universe. Biologists tell us that creation of life is an incredibly improbable event. Nevertheless, the probability is non-zero, since we are here. Most exobiologists agree on this fact, if on little else. The probability of there being any intelligent extraterrestrial life is necessarily even smaller. Attempts have been made to quantify these statements. Vast sums are being expended on searches for traces of alien life. Yet there is one overwhelming pre-eminent fact, namely that our ignorance of the origin of life prevents us from coming to any definitive conclusions about the pervasiveness of life.

The Dominican philosopher friar Giordano Bruno argued that the universe was teeming with extraterrestrial species. He was burnt in Rome in 1600 for his unorthodox views. The US physicist Frank Tipler took the seductive step of pointing out that if anywhere else in our galaxy there was a planet harbouring an intelligent civilization, the inevitable consequences would lead to discovery. He argued that over the billions of years available to such a civilization to develop technological maturity, it would develop the techniques for sending self-propagating robotic probes on a task of exploration of our galaxy. These probes would have certainly encountered the Earth and solar system, and left visible traces for us to discover. He claims that the universe cannot be teeming with life, because if it were we would have already found artefacts of ancient alien civilizations in the solar system. One problem with this argument is that it ignores the likelihood that any advanced civilization capable of colonizing the galaxy would have developed the ability to hoodwink and hide from any terrestrial simpletons that they encountered.

Tipler's argument was based on the premise that in our galaxy there are about one hundred billion solar-type stars, each of which is likely to have a planetary system resembling our own. Astronomers have taken the first steps towards discovering such planets, although so far modern technology has only led to the detection of very massive Jupiter-like bodies. These are almost certainly inhospitable to life. However, these massive planets are just the tip of the iceberg of the swarm of bodies that are orbiting distant stars. Tipler argued that life was a likely occurrence, over the billions of years of available time.

This need not be the case, however. In fact, estimates of the probability of life that stems from biological facts range from infinitesimal to negligible, relative at least to the number of planets in our galaxy. The US astronomer Frank Drake attempted to by-pass this uncertainty by developing in 1961 an equation that has played a role in motivating ongoing searches for extraterrestrial intelligence.

Drake's equation consists of the following. Take the number of Sun-like stars. This is about one hundred million. Multiply by the number of Earth-like planets per star. This is likely to be about one, or perhaps 0.1 if only one candidate star in ten has such a planet. Multiply by the probability that life originated on this planet. Here is where the mystery first enters. Biology cannot yet account for the origin of life. Some element of self-organization is needed. After all, this worked at least once, despite the odds. Optimists would say that given the right conditions, namely water, clay, and a mild climate, life inevitably develops from this primordial ooze. Let us remain optimistic, and say that 1 per cent of such Earth-like planets may spawn life. One can then allow for the fact the evolution inevitably, or say at least 10 per cent of the time, generates intelligent life, and that over the ten billion years of the Milky Way galaxy, there is a reasonable chance, say 1 per cent, that such life survives the various crises that would have inevitably occurred. We are left with a hundred thousand alien civilizations in the galaxy, vastly advanced relative to the inhabitants of planet Earth. The newest such civilization is a mere one hundred light years away.

Such odds would seem to make it well worth developing the eavesdropping technology that could enable us to listen to our nearest neighbour's electronics traffic. This has been the goal that has stimulated several ongoing experiments designed to intercept such signals. Once transmitted, electronic signals propagate endlessly through space. One could detect the Earth at a distance of up to several light years, since only seventy years have elapsed since radio broadcasting was established. Alien civilizations with an advance on the Earth could be detectable millions of light years away were our radio telescopes sufficiently sensitive.

Government funding has not favoured searches for extra-terrestrial intelligence. Two of the major experiments under way have found private benefactors. Film-maker Steven Spielberg and Microsoft co-founder Paul Allen have independently aided the development of radio receivers with millions of channels, which are one of the crucial ingredients in detecting and deciphering weak radio signals for any evidence of an intelligent origin. The dream behind such efforts is that while the odds for success are small, any successful detection would have immense implications for humanity.

The probability of life

Unfortunately there is little substance to estimates of the number of intelligent civilizations. The reason is simple. One of the factors in Drake's equation requires an estimate of the probability of life. Not only is this probability unknown but also the biologist's best estimates of such an occurrence range from the truly infinitesimal to a number far smaller than one hundred billion. With a number as small as the most optimistic of quantitative estimates based on biology, one would not expect a single life form to have emerged among the supply of possible planetary sites.

This assumes that random sequences of events were the precursor to life. Such estimates are based on random arrangements of molecules arising spontaneously to provide primitive proteins and chains of DNA. However, there are many examples in nature that demonstrate that one can overcome the unlikeliness of spontaneous events arising from random circumstances with a certain degree of self-organization. Weather patterns demonstrate how minor changes in humidity or pressure can precipitate dramatic changes. Critical phenomena are commonly encountered where minor perturbations have great consequences, as in, for example, the onset of fluid turbulence or the phenomenon of superconductivity. The flutter of a butterfly's wings in China can induce a tempest halfway across the globe, so proponents of complexity theory tell us.

Exobiologists hope that such phenomena may enhance the probability of the occurrence of life. In the absence of any facts, however, there is little input for such expectations, apart from the single indisputable fact that life once emerged on one planet. The only reasonable inference is that the probability of the emergence of life was infinitesimal, and far smaller than would allow one reasonably to expect to encounter another civilization in our galaxy. But it was not completely zero, otherwise we would not be here.

There are some ten billion galaxies in the observable universe, each with up to a hundred billion terrestrial-like planets. If one could somehow surmount the challenge of voyaging through intergalactic space, something like a billion trillion Earth-like planets become accessible. But if the odds of spawning life are as infinitesimal as our knowledge of nature suggests, then even a billion trillion planets would not guarantee emergence of a life form.

Recent developments in astronomy suggest that the outlook for the existence of alien life may not be completely bleak. Evidence has accumulated that the universe is infinite, or at least very much larger than the observable extent. If this is true, there may be a vastly larger number of planets, and possibly an infinite number. In an infinite universe, there are an infinite number of life-sustaining planets, and an infinite number of life-generating events will have occurred. Life, however improbable, must inevitably have been spawned somewhere. Moreover, one could logically expect there to be life forms vastly superior in intelligence and technology to our own.

Most typically, each of these life-spawning events will have had some ten billion years or so to evolve and hence achieve a state of intelligence and technology superior to ours by several billion years. Even in a finite but very large universe, the odds could be favourable for a species to exist somewhere of vastly superior intelligence to our own. In an infinite universe, there is no doubt whatsoever that this must have happened long ago.

In the thousands of years of cultural and scientific life on Earth, mankind has made immense progress. Yet one cannot conceive of the immense potential of a civilization that has been flourishing for a billion, or even a million, years. Our achievements are likely to have as much impact on such world citizens as an ant colony has on us.

If the universe is infinite, someone somewhere may have unlocked the secrets of time travel. It is impossible to imagine the state of superiority such advanced civilizations might have achieved, if they survived. At the very least, we must concede that they would have mastered wormhole technology. At the same time, however, most of the volume of the universe is inaccessible to us. A radio signal emitted by us would only have travelled a few tens of light years. Some hypothetical highly advanced species could at most have explored a domain a few billion light years across since the Big Bang, which is known to have occurred 14 billion years ago. But the universe is vastly larger than this, and the likelihood is that any advanced cosmic cohorts would be billions of billions of light years away. Only at such a great distance would one be likely to have enough planets for at least one to have generated life with reasonable certitude, given the low odds for life.

And intelligent life?

> The idea that we shall be welcomed as new members into the galactic community is as unlikely as the idea that the oyster will be welcomed as a new member into the human community. We're probably not even edible.
>
> John Ball

There is an interesting statistical argument one can give that runs somewhat counter to the optimistic tone of the preceding discussion. Intelligent life developed in a narrow window of opportunity that may have lasted no more then a few hundred million years. One had to have a suitably calm and benevolent host star, and this takes several billion years in order to synthesize the heavy elements. Of course we have but the one example, but it would seem plausible to generalize to the average timescale for the evolution of intelligence anywhere in the universe.

Nor can we afford to wait too long, since a planet would be uninhabitable once its parent star evolved off the main sequence. It would not be much fun to inhabit a planet orbiting a red giant star. The implication from this is that if the window of opportunity is narrow, intelligent life is expected to be but a rare occurrence in the universe.

Conversely, primitive life forms should be ubiquitous, for the simple reason that on Earth the earliest life forms evolved relatively rapidly. The Earth itself formed about 4.6 billion years ago. Life has existed on Earth for at least four billion years. Life probably could not have existed any earlier. The first half a billion years on Earth presented a desperately hostile environment for biogenesis. During this period, there was no protective atmosphere and meteoroids unceasingly bombarded the surface of the young Earth. Indeed, one particularly massive impact led to the formation of the Moon.

As pointed out by astrophysicists Charles Lineweaver and Tamara Davis of the University of New South Wales, this confluence of violence with the evidence for primitive life forms shows that biogenesis must have occurred within a very short time on the cosmological scale. These authors estimate that this time, bounded by the period between the last sterilizing major impact and the oldest evidence for life on Earth, was most likely to be as short as a hundred thousand years, and no longer that half a billion years. The inference is inevitable that, statistically speaking, simple life forms abound throughout the universe. The step that led to the transition from primitive to intelligent life forms is likely, however, to have been a rare and perhaps even a unique event. This inference should certainly give comfort to the theologians.

Where are they?

> Sometimes I think we're alone. Sometimes I think we're not. In either case, the thought is staggering.
>
> Buckminster Fuller

This leads us to a paradox first posed by Enrico Fermi: 'If they existed, they would be here.' If the universe is teeming with advanced species, why haven't we encountered them yet? The answer must be that such a high threshold is required to master wormhole technology that any voyaging civilization would be capable of taking thorough measures to remove any traces of their visitations. Surely they could observe us without our being aware of being under examination. Modern spying technology is already capable of this, and one cannot really conceive of the sophistication of spying techniques that would become available a century hence, let alone a billion years from now.

Would they actually have done so? In so far as they might regard us much as we like to imagine primitive tribes in Amazonia, it would seem entirely plausible that we would be encouraged to continue in ignorance of any superior technology that was surreptitiously surveying us. There need be no paradox in that no evidence of any alien artefacts have been found in the solar system. Any superior civilization that has developed wormhole technology would be sufficiently sophisticated to be able to hide all traces of its voyages.

There is one trace it may have left, however, and that is the spark that ignited life on Earth. This could constitute a controlled experiment that had a successful outcome. Perhaps there were many others that failed, or that succeeded. If the possibility exists of external triggering of life, then we return to our initial dilemma. We cannot calculate the probability of the existence of a nearby advanced civilization, for the rules of the game are unknown. We must search.

Travelling in time

> If you do not ask me what is time, I know it; when you ask me, I cannot tell it.
>
> St Augustine

> Everything that you could possibly imagine, you will find that nature has been there before you.
>
> John Berrill

Even if intelligent life is rare, the universe is vast. Imagine, for example, that the next species worth contacting were so remote that it was not in direct causal contact with our solar system. It could have developed in some remote part of the universe so far away that light would not have had time to travel from there to us. The possibility for any interaction would seem to defy the laws of physics. This is not necessarily the case, however. Physicists can, in principle, negotiate their way around space and time via novel technologies. There need be no barriers to exploring the future, although we shall see that the past – that is, the past before the epoch of time travel – is inviolate.

An infinite, or nearly infinite, universe offers interesting possibilities for the would-be time-traveller. The essence of time travel is the existence of trajectories that loop back in time. A spacecraft that steered along such a

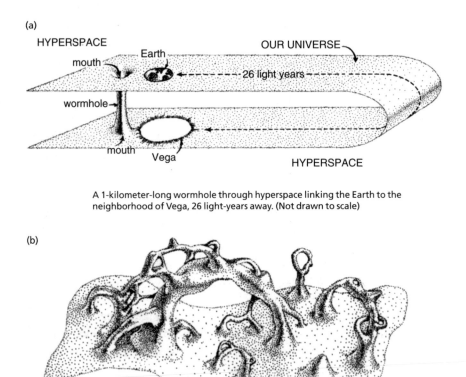

(a)

HYPERSPACE   OUR UNIVERSE

mouth   Earth

← - - - - - 26 light years - - - - - - - →

wormhole →

mouth   Vega   HYPERSPACE

A 1-kilometer-long wormhole through hyperspace linking the Earth to the neighborhood of Vega, 26 light-years away. (Not drawn to scale)

(b)

Figure 11 (a) A worm hole trajectory linking the Earth to Vega (schematic). (b) Space is full of (virtual) worm holes.

Source: K. S. Thorne (1994) *Black Holes and Time Warps*. New York: Norton.

path, called a closed time-like line, would follow a trajectory that took it into the past or the future. Closed time-like lines are not forbidden by Einstein's theory of gravity. Perhaps they therefore exist. Indeed, there is a notable belief held by physicists that anything that can exist must exist, or must have existed. This justifies their hitherto fruitless searches for such exotica as primordial black holes, fractionally charged particles, or even particles, called tachyons, that travel faster than light.

There are other exotic possibilities for time travel. Time machines in the form of wormholes are a part of modern physics. A wormhole is a potential tunnel from the inside of one black hole to another. Wormholes as well as black holes are a topic beloved by science fiction writers. A space–time traveller can emerge in the past by virtue of travelling along a closed time-like loop. If wormholes exist, so do closed time-like loops. The world's scientific authorities on time machines prefer to discuss closed time-like loops in order to hinder journalistic recognition of what they are really discussing. This endeavour has largely been successful, at least in the short term. Worm holes and closed time-like loops are now an accepted part of general relativity theory.

The quantum theory has changed our outlook on wormholes. Wormholes exist in the world of virtual reality. Of course, whether wormholes actually exist or not is another issue, and their existence has even been deemed by some to be incompatible with physics. In the words of the British cosmologist John Barrow, 'it is easy to assume that the inclusion of wormholes is too radical a step to take in enlarging the picture we have of space and time, but it is equally possible that it is a step that is not radical enough'. However, once their existence is allowed it would seem churlish to deny their existence in the absence of any experimental proof.

However, wormholes rapidly self-destruct, and we never actually notice their fleeting existence. To make use of them, we need to invoke a technology from the remote future. The most difficult step is to trap a wormhole and prop it open for use by intrepid would-be time travellers. There is no law of physics that forbids such manoeuvring. This requires appeal to the same principle by which black holes radiate, where the intense gravity force near the black hole can effectively separate virtual particle pairs. Once separated, one of the particles can fly off to infinity and thereby drain energy from a black hole. No one has actually detected this effect, but physicists are confident of the principle that black holes actually

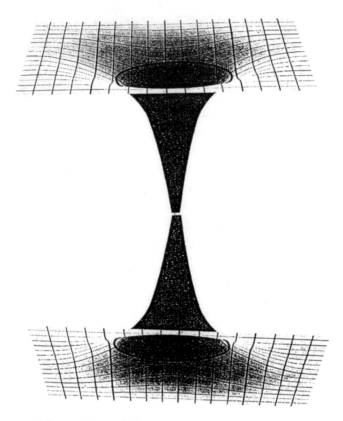

Figure 12 Black hole and wormhole.

radiate, as first proposed by Stephen Hawking. If this works for black holes, then it is not much of a stretch to imagine that wormholes in existence for the merest instant can be gravitationally disrupted using the strong gravity field near a black hole to enable wormholes to be extracted from the vacuum.

The theory of time machines is all very well, but what ultimately counts is reality. Wormholes provide the engineer's blueprint for constructing a time machine. If black holes exist, then wormholes almost certainly must exist. Whether they are physically accessible, and whether one can regulate and control their formation, are quite different, and essential, matters. Mastery of wormhole technology is needed to construct a time machine. Wormholes were, and are, a fad that some brave souls thought might contain the key to unlocking the ultimate secrets of nature. Unfortunately, to reach a wormhole, one has to penetrate inside the event horizon of a black hole. This is a difficult task, as any conventional material is ripped to shreds by the overwhelming tidal forces as one approaches the horizon.

The implications of the existence of such travel are remarkable. The Earth would no longer be immune from alien contact. Once wormhole exploration of space–time is feasible, one can readily imagine the development of self-propagating probes. By going to the distant future, all parts of the universe are accessible, including those now beyond our horizon. By going into the past, the probes could monitor or stimulate evolution. Wormhole travel is essentially instantaneous. There are no boundaries in space or time. Once wormhole technology is available, all parts of the infinite universe are accessible, and could certainly be explored by a single species originating at a unique point in space and time.

17 A Brief Moment in Time

From a wild weird clime that lieth, sublime
Out of Space – out of Time

Edgar Allan Poe

Things are always at their best in their beginning.

Blaise Pascal

The past is unique and inviolate from the wanderings of a future time traveller. It is not possible to travel back to an epoch where no time machine already existed. A time traveller needs an exit door as well as an entrance: for this, time machines are needed. Once some future civilization masters the techniques of time machines, from then onwards what we think of as space and time will never be the same again.

We can imagine the equivalent of super highways laid down in time as well as in space once time machine construction becomes routine engineering. The future will be multiply connected. Only the past is sacrosanct. We need not worry about time travellers appearing at our front door in the immediate future. Shakespeare can rest in peace: there is no danger that his greatest plays were written by some traveller from the future.

There is one notable exception about travelling into the past. Exotic physics may be present near a black hole. Naturally occurring wormholes may be lurking inside the black hole, available to the intrepid explorer who is willing to risk all for the voyage. The traveller disappears from our universe: all communication is lost. In her frame of reference, however, new vistas and new dangers beckon. The danger is bombardment by highly energetic gamma rays that are the residue of the highly redshifted light that leaked into the black hole from the external universe, and subsequently became highly blueshifted. Such radiation would threaten the existence of any creature. But let us imagine this poses no problem.

Into the singularity

There are even more exotic objects awaiting discovery, on the principle that anything permitted to exist by the laws of physics must exist, at least somewhere in the universe. Pride of place must go to the complete breakdown of space and time that happens at the centres of black holes. Normally one cannot get close enough to inspect such a point in space–time, thanks to the protective existence of a horizon. Event horizons are insurmountable barriers to the intrepid space traveller. Too close an approach results in destruction by immense tidal gravitational fields. But Einstein's theory of gravitation allows, in principle, the independent existence of such singular points, unveiled by any horizon. They are called naked singularities. Anything can happen at a naked singularity. Cause and effect lose any meaning. There is no future, there is no past.

The US physicist Kip Thorne advocates the use of naked singularities as a means of achieving time travel. These are horizon-free and exceedingly hard to find. There is a conjecture due to Roger Penrose that naked singularities are forbidden to exist, despite the permissiveness of Einstein's equations. Such cosmic censorship need not frustrate the dedicated time traveller. Appealing to the uncertainty principle of Heisenberg, she notes that wormholes, and naked singularities into the bargain, may come and go out of thin air, actually a perfect vacuum, for brief instants of time. Over the Planck time, a mere 10^{-43} seconds later, the wormhole closes. No laws of nature are violated if one borrows energy or mass from the vacuum for a sufficiently short, immeasurably short, instant of time, provided that the loan is promptly repaid. Our next goal is to keep an emerging wormhole propped open. This can be done with exotic material, whose energy density (rest mass plus pressure contribution) is negative. Never mind that mere mortals, and astronomers, have never encountered such stuff. The laws of physics do not forbid it, and that is enough for the keen adventurer to leap forward, down the wormhole. She emerges at some point in space and time, far in the future, and perhaps in the past. In this way one can voyage throughout an infinite space, and enter into universes that are otherwise inaccessible.

The central singularity beckons, and offers the possibility of an entry into another universe. In practice this means some arbitrary displacement in space and time, outside the horizon of the region of the universe accessible by observers impatiently tracking the fate of the time traveller from a safe vantage point outside the black hole. No route planning is possible. The journey would be hazardous and random, and subject to the laws of quantum uncertainty that must apply on scales close to the singularity. This voyage would not pose any challenge to the uniqueness of the past of any observer. Any particular trajectory would be effectively unperturbed by future travellers: the risks would be negligible.

For the citizen of the future, the prospects might at first seem boring. In the epoch of time travel, there would be little scope for free will or creativity. Repeated trips in time by denizens of the far future would reshape the past. No future poet could ever be confident that his inspiration had not come from the future.

Consider a pebble on the beach. A time traveller arrives and pushes the pebble to the right. He leaves for the future, and then returns to the epoch just before he displaced the pebble, and moves the pebble to the left. Where does the pebble end up? Does this question even make sense? The position of the pebble is dependent on the future as well as the past. Free will cannot really exist for the prospective pusher of the pebble: he has little say in its location.

It gets worse. You write a great novel. Or so you are convinced. But did you really create the novel? Perhaps a time traveller arrived from the future and telepathically or hypnotically provided you with the story. Or even typed it in directly to your computer while you were sleeping. It is hard to disentangle the facts. Indeed, are there any unique facts at a specified time or place? The answer must be no: space and time have become inextricably entangled.

The fabric of space and time has become interwoven with itself. There is a unique time but the properties of space at this time depend on other times, in the future as well as in the past. A walk in space has become a journey through time. The converse is also true: what one thinks of as time becomes space. One instant in time may connect remote parts of space.

Curiously, we know that the intermixing of space and time occurs near a black hole. It is a prediction of Einstein's theory of gravitation. An observer who passes too close to a black hole will lose all contact with the outside world. Gravity will draw him ever closer, as all contact is lost with any external observers. Any signals he sends out are increasingly redshifted into invisibility. Time goes on in the outside world, but time slows inexorably for the black hole traveller as seen by the external observer. He disappears from view. There is a difference, however: space opens up as externally viewed time grinds to a halt. He falls into the central singularity. In his own reference frame, time continues to tick. What happens next is anyone's guess. Re-emerging far away in space and time is one possibility. Time travel would be achieved in this case.

The dawning of the epoch of time travel would be equivalent to a voyager's approach to a black hole. The local fabric of space and time would have to change. A physicist's success in constructing a time machine would have changed the fundamental structure of the universe.

This is an example of a phase transition. Such transitions are an essential part of our understanding of the universe, albeit that the time traveller's adventures would spawn something that truly has no precedent. More mundane phase transitions occur every day. The freezing of water is an example of how the local molecular structure can undergo a sudden change, everywhere, at once, once the temperature falls below the freezing point. Cosmology provides another example. As we have seen, the period of inflation, lasting just 10^{-35} seconds, was such a phase transition from which our universe emerged.

18 Into the Infinite Future

The world's a bubble, and the life of man
Less than a span.

Francis Bacon

So many worlds, so much to do,
So little done, such things to be.

Alfred, Lord Tennyson

It is apparent that the matter density is below the critical value required to halt the expansion. Utilizing the current value of Hubble's constant, one finds that the critical density is 10^{-29} grams per cubic centimeter. This is equivalent to 0.00001 hydrogen atoms per cubic centimetre. This may seem a small value, but we have already seen that the density of luminous matter, when averaged over all of space, is a negligible fraction of the critical density. Since the total density of dark and luminous matter amounts to a third of the critical density, the universe is destined to be always dominated by its kinetic energy: it will expand forever. This is because if an expanding shell is not being decelerated by the effect of the gravity of matter in its interior, its gravitational potential will decrease as the radius of the shell increases. The gravitational potential energy forms a smaller and smaller fraction of the kinetic energy as the shell expands. The universe will become a colder and darker place as galaxies recede from one another.

It matters little that the presently observed universe may be finite. Suppose it turns out that the curvature of space is flat but that the universe is finite. The observations of cosmic acceleration from distant supernovae show that it will expand forever. In the distant future, one would then have an almost infinite number of planets and an almost infinite time for life to develop. Planets will survive until protons decay, at least 10^{32} years from now. These will not be planets like our own Earth in orbit around a yellow dwarf star. While the number of currently existing planetary sites is limited, over such a vast timescale many possibilities arise for forming new planets as stars and galaxies evolve. The environments will be exotic; for example, the future evolution will mostly occur in the dense cores of galaxies. But in these regions, where dynamical evolution is currently rapid, supermassive black holes are found. These black holes provide the ultimate energy source, as matter accretes on to them, and will continue to light up the universe for countless eons. Planets are present in highly evolved environments; for example, around neutron stars. As long as there are black holes, one can imagine a source of energy for keeping planets warm and available for potentially seeding life. Given enough time, even a finite number of sites will allow the likelihood of life.

The future of the universe is set to be a long, drawn-out process. The universe that astronomers observe is but a minute grain on the way towards the future extent of our horizon. The Big Bang theory enables us to reconstruct the brief flash of brilliance, minutes in duration, when the light elements were synthesized. Our theory of the evolution of stars describes how all the other elements were made. All of this happened within the past ten billion years. Stars will continue to shine via nuclear fusion of hydrogen until the universe has an age of ten thousand billion years. After that, the universe will consist of degenerate stars, such as brown dwarfs, white dwarfs, neutron stars, and black holes. It will be a dark and diffuse place, where there is no longer any radiation from ordinary stars to light up the night skies or warm the planets.

The ultimate fate of the universe

> The most important fundamental laws and facts of physical science have all been discovered, and these are now so firmly established that the possibility of their ever being supplemented in consequence of new discoveries is exceedingly remote.
>
> Albert Michelson

This is as far as sound speculation takes us. But it is possible to extract the

Figure 13 (a) In the old Big Bang theory, the universe is expected to have a size that is not much larger than we can already probe with the largest telescopes.

feeblest flickers of possible life forms in an increasingly desolate environment. Degenerate stars live on thermal energy that is radiated away. Their thermal energy source can occasionally be revived if new degenerate stars form, as would happen when mergers occur between pairs of orbiting white dwarfs. Planets can even form in the gaseous disks that inevitably develop, thanks to conservation of angular momentum. A faint glimmer of energy from degenerate stars continues as weakly interacting dark matter particles (if they are massive as theory suggests) are trapped in

Figure 13 (b) In the inflationary universe, the size of the universe is vastly larger than we can expect to observe in the foreseeable future.

Source: E. Mallove (1988) 'The self-reproducing universe'. *Sky and Telescope,* September.

stellar cores, where they accumulate and annihilate, releasing gamma rays and energetic particles that heat the star.

The next era of the universe is demarcated by proton decay. This is a consequence of the theory of unification of the fundamental forces, although there is as yet no experimental confirmation. Protons rapidly decay and are created at the energy scale of grand unification (10^{16} GeV), and this requires that proton decay must also occur at low energies, but very slowly.

There is no immediate danger to us, since there has not been enough time in the universe for many decays to occur. It is interesting that in the human body, which contains about 100 kilograms or about 10^{29} protons, there could be approximately one proton decaying per decade. The resulting gamma ray flux is not enough to produce any risk of cancer. Had the proton lifetime been as short as 10^{14} years, though, we would all be dropping dead of cancer.

Although slow, the rate of proton decay predicted by grand unification is measurable. Several experiments to detect proton decays have been undertaken. The principle is simple: take 1000 tonnes of water containing about 10^{32} protons, and, on average, ten protons should decay per year in the simplest versions of grand unification. Since the decay of a proton releases a lot of energy (in the form of energetic muons and gamma rays), this effect can be detected, at least in principle. However, one has to shield the experiment from cosmic rays that induce muon showers in the Earth's atmosphere. A typical experiment uses tens of thousands of tonnes of purified water stored in vats deep underground, surrounded by scintillation counters that search for light flashes from the rare decay events. To date, no protons have been observed to decay. From the absence of proton decay, it is inferred that the lifetime for the proton exceeds 10^{32} years. Diamonds could be virtually forever, if this is any consolation. For elementary particle theory, the conclusion, for the moment, is that the simplest models of grand unification are wrong.

Nevertheless, after 10^{50} years have elapsed, one can be reasonably confident that all protons in the universe have decayed. The theory of gravitation requires this. Protons are swallowed up by virtual mini black holes and decay into positrons. The only objects left are black holes.

Even black holes do not last forever. Stephen Hawking predicted that the tidal forces near the event horizon of a black hole are so strong that pairs of virtual particles, which constitute the fabric of space and time, can be ripped apart. Virtual particles exist for a time so brief that the uncertainty principle tells us that conservation of energy is not violated, since these particles are unobservable. The extreme curvature of space near a black hole is able to separate the virtual pairs into real particles, one of which escapes to infinity and thereby provides a channel for the black hole to evaporate. The smaller black holes will be the first to self-destruct, but eventually, after 10^{100} years, even the most massive black holes will have evaporated. The universe at this point consists of the leftover waste from stars: radiation of very low frequency, neutrinos, and positrons and electrons.

On immortality

Can one imagine intelligent life continuing forever? We can dissect intelligence into computability, taking a suitably ascetic approach that ignores, say, emotion or passion, art or poetry. The indefinite continuation and progress of intelligence then requires us to imagine a potentially infinite, or certainly unlimited, number of computations. Life can be regarded as infinite if the amount of energy generated is finite. Otherwise one's brains would be baked, if brains were the appropriate storage devices for intelligence.

Now computations generate heat and require dissipation of energy. Freeman Dyson pointed out that one needs a minimal operating temperature for any organism, as otherwise it cannot dissipate the heat it generates and simply fries itself to death. This minimum temperature is very low, about a trillionth of a kelvin, and rather far from any immediate challenge to society despite the ever-cooling expanding universe. However, one day, life would have to grind to a halt given the inexorable expansion of the universe. Dyson's strategy to evade this heat death was to suggest that life could find a way to hibernate. Life need only radiate energy intermittently. In this situation it turns out that in an expanding universe at the critical density, life can be considered infinite, since the energy dissipated is finite.

However, a crisis for life has emerged. Recall that the universe is considered to be full of the dark energy that is responsible for the observed acceleration. In the presence of the dark energy associated with the cosmological constant, the vacuum energy is eventually the dominant energy form. There is a thermal bath of virtual particles that uniformly fill space. The possible channels for a computing, living system to radiate become blocked, as there are no thermodynamic possibilities for dissipation of energy to occur. Life systems overheat, life is over.

This would seem to be an unfortunate consequence of the concordance model of cosmology to which many astronomical observations are pointing. There are two possible ways out. The mystery of the cosmological constant can be summarized as: why now? It is only recently emerging as the dominant energy source in the universe. Long ago – for example, at the epoch of inflation – the cosmological constant term was negligibly small. Its energy density was less than that of the radiation content by more than a hundred powers of ten.

This seems a remarkable piece of fine-tuning of the initial conditions of the universe, and a counter-proposal has emerged. This has been dubbed quintessence, a new component of dark energy, which is designed so that it always tracks the dominant component of the universe. In other words we have a cosmological constant that is no longer constant, but decays with time. Not only can such a postulate sidestep the need for fine-tuning, because inflation itself is driven by a cosmological-constant-like force field, whose strength, however, is comparable to the radiation density at that time, and vastly larger than the cosmological constant measured today. Because the cosmological constant will continue to decay, there is no longer any barrier to life systems continuing to radiate. Life may be forever. Of course, any finite system is only capable of storing a finite number of memories. As pointed out by physicists Katherine Freese and William Kinney, we have to be reconciled to the fact that 'while life itself may be immortal, any individual is doomed to mortality'.

In the absence of quintessence, there is not much prospect for immortality. Even in a cyclic universe, one might expect that life would be renewed in each cycle. However, one would dearly like to know whether any memories might make it through the singularity that separates the contraction and re-expansion phases. Physics has no answer.

Wormhole technology might help in enabling us to tap new resources of energy for our computational needs. Or one could just pack up and say goodbye, resettling in a warmer, more congenial climate in some time and place beyond the horizon.

The end as a new beginning

The universe could be destined for boring senescence, but more extra-ordinary possibilities may await the intrepid space–time explorer.

University of Michigan physicists Fred Adams and Greg Laughlin have proposed one of the more remarkable of these. The possibility exists of a phase transition that may, literally, open new horizons.

The present universe acquired its characteristics via several phase transitions. The most noteworthy of these, now generally accepted by most cosmologists, was that of cosmic inflation. The universe was caught in a state of perfect symmetry as it cooled after the Big Bang. This is not good: our universe is very definitely far from symmetric, as otherwise we would not be here. Matter dominates over antimatter by at least ten thousand parts to one. But quantum physics permits the universe spontaneously to exit the symmetrical state. In the symmetrical state, the vacuum has a large energy density. We call this a false vacuum. The universe quantum tunnelled its way into the true vacuum. During the brief period when the energy of the false vacuum was dominant at about 10^{-35} seconds after the Big Bang, the universe underwent the phase of exponential expansion known as inflation. This is how one understands the present size of the universe and indeed the origin of its structure via seeding of inflated quantum fluctuations.

The vacuum at present is characterized by a very low energy density. But a new form of energy could be lurking in the vacuum, and be capable of triggering a future phase transition to a new vacuum state. Such a transition is capable of spontaneously spawning exponentially inflating bubbles of new phase. The bubbles have kinetic energy. As bubbles crash together, reheating of the universe occurs. In effect, one has created new universes. Some cosmologists even speculate that our universe is simply the largest and most dominant of these secondary universes. The long-term future of the universe might be rebirth.

19 And so to God

By God I mean a being absolutely infinite – that is, a substance consisting in infinite attributes, of which each expresses eternal and infinite essentiality.

Baruch Spinoza

I believe that a scientist looking at non-scientific problems is just as dumb as the next guy . . .

Richard Feynman

The role of God in cosmology has both intrigued and plagued cosmologists since the earliest recorded writings on the nature of the cosmos. Some suffered: not only were their books banned or burnt, even they were not immune from similar treatment. In modern times, the Catholic Church at least developed more progressive views under the guidance of the eminent Jesuit cosmologist Georges Lemaître.

Born in 1894 at Charleroi, Belgium, Lemaître, as a young, unknown researcher and already ordained, was able to demonstrate in 1920 that Einstein's static universe from 1917 was unstable to collapse. In 1927, Lemaître proposed that the universe was expanding in order to take account of the accumulating data of Vesto Slipher and Edwin Hubble, which demonstrated that the galaxies were receding from us. Einstein did not take kindly to Lemaître's work on cosmology, qualifying it as '*physique de curé*'.

All changed in 1929, however, when Hubble announced the eponymous law that stated that the recession velocity of a galaxy, as inferred from its redshift, is directly proportional to its distance from us. Lemaître's explanation was that space is expanding, carrying along the galaxies like so many specks of dust in a wind. This represented a revolutionary step forwards in our understanding of the universe that Einstein soon accepted. Our view of the universe had changed irrevocably.

Lemaître was readily able to reconcile the Big Bang cosmology with his

Catholic faith. Indeed, his advice played an important role in the famous encyclical, Humani Generis, that Pope Pius XII issued in August 1950. The wording was discrete but firm.

If anyone examines the state of affairs outside the Christian fold, he will easily discover the principal trends that not a few learned men are following. Some imprudently and indiscreetly hold that evolution, which has not been fully proved even in the domain of natural sciences, explains the origin of all this, and audaciously support the monistic and pantheistic opinion that the world is in continual evolution. . . . Catholic theologians and philosophers, whose grave duty it is to defend natural and supernatural truth and instill it in the hearts of men, cannot afford to ignore or neglect these more or less erroneous opinions. Rather they must come to understand these same theories well, both because diseases are not properly treated unless they are rightly diagnosed, and because sometimes even in these false theories a certain amount of truth is contained, and, finally because these theories provoke more subtle discussion and evaluation of philosophical and theological truths.

The spiritual soul of man remained sacrosanct. But Humani Generis is a milestone in the evolution of Catholic thought on evolution, for it cleared the way for an allegorical interpretation of the Genesis account of the creation of the Earth, the heavens and humanity.

On Lemaître's deathbed in 1966, he heard about the discovery a year earlier of the relic radiation from the Big Bang, the cosmic microwave background radiation. This was a fossil that he had long sought, convinced of its existence, albeit under the very different guise of high-energy cosmic rays. To the end, Lemaître's preferred model of cosmology, now known to be mistaken, was that of a primordial massive atomic nucleus, decaying explosively to form the expanding universe. His vision of the past glories of the universe is vividly described in his writings: 'Standing on a well-chilled cinder, we see the fading of the suns, and try to recall the vanished brilliance of the origin of the worlds.' Lemaître was uniquely able to reconcile religious beliefs with cosmology, best summarized in his own words penned in 1950:

L'Univers n'est pas hors de la portée de l'homme. C'est l'Eden, c'est ce jardin qui a été mis a la disposition de l'homme pour qu'il le cultive, pour qu'il le regarde. L'Univers n'est pas trop grand pour l'homme, il n'excède pas les possibilités de la science, ni la capacité de l'esprit humain.

(The universe is not beyond the realm of man to comprehend. It is his Eden, his garden to cultivate and to observe. The universe is not too vast for man, and for the possibilities of science and of the human spirit.)

The Vatican's position continued to evolve. In the words of Pope John Paul II in October 1996 during an address to the annual meeting of the Pontifical Academy of Science, there was no longer 'any opposition between evolution and the doctrine of the faith about man and his vocation'. Evolution had finally become theologically respectable.

Cosmology may have become theologically acceptable. However, the jury remains out on Darwin, at least as far as the numerous proponents of non-random design are concerned. Many refer to the hand of God in evolution by invoking intelligent design, capitalized or not, although the role of intelligence is a matter of belief.

God in the past

Design arguments were often used in the eighteenth and nineteenth centuries to evoke God. The design of the universe, to some, is a manifestation of the work of God. The British poet Percy Bysshe Shelley eloquently expressed the theological implications: 'I think that the leaf of a tree, the meanest insect on which we trample, are in themselves arguments more conclusive than any which can be adduced that some vast intellect animates Infinity.' This is by no means a consideration restricted to modern times. Consider Marcus Tullius Cicero: 'The celestial order and the beauty of the universe compel me to admit that there is some excellent and eternal Being, who deserves the respect and homage of men.' Bernardin de Saint-Pierre (1773–1814), a French naturalist and priest, was fond of explaining how nature was designed for man's well-being. To him, the melon was divided into slices by Nature in order to be eaten by a large family, and the much larger pumpkin was meant to be shared with the neighbours. Nor is one confined to material aspects. With Friedrich von Schlegel, it is but a short step from concepts to divinity: 'Ideas are infinite, original, and lively divine thoughts.'

Darwin showed how the appearance of design can arise in living things through the process of natural selection. The remaining place for God was pushed back to that of the originator of the cosmos as a whole. It does not take much of a step to infer that the universe has an intricate and beautiful design that requires the hand of a Grand Designer. Not all authors agree, however, as in the case of Arthur C. Clarke: 'If there are any gods whose chief concern is man, they cannot be very important gods.'

Others see infinity as the mark of God. And infinity may be buried, perhaps in poetry. Indeed, the literary world, notably Jean Cocteau, is attracted to infinity: 'Mystery has its own mysteries, and there are gods above gods. We have ours, they have theirs. That is what is known as infinity.' But it is not clear to the British philosopher and mathematician Bertrand Russell that infinity has any uniform meaning or definition: 'If any philosopher had been asked for a definition of infinity, he might have produced some unintelligible rigmarole, but he would certainly not have been able to give a definition that had any meaning at all.' The lack of a definition, of course, is no reason to stop looking for the meaning of infinity. Infinity, nevertheless, as revealed by nature, may be beyond the realm of science. Some great physicists, such as Max Planck, subscribed to this view: 'Science cannot solve the ultimate mystery of nature because we ourselves are part of nature and therefore part of the mystery we are trying to solve.'

Philosophers as well as theologians are capable of seeing God everywhere, and especially in the workings of the universe. According to Immanuel Kant, 'God has put a secret out into the forces of Nature so as to enable it to fashion itself out of chaos into a perfect world system.' Some of the most eminent scientists are closet theologians. Even for Albert Einstein, 'Science without religion is lame, religion without science is blind.' And more: 'That deeply emotional conviction of the presence of a superior reasoning power, which is revealed in the incomprehensible universe, forms my idea of God.'

Other scientists generally have little patience for philosophers who theorize about God. Thomas Huxley was a leading proponent of this viewpoint:

Of all the senseless babble I have ever had occasion to read, the demonstrations of these philosophers who undertake to tell us all about the nature of God would be the worst, if they were not surpassed by the still greater absurdities of the philosophers who try to prove that there is no God.

Yet others seem to be willing to follow the lead of Blaise Pascal. If God exists, the believers have gambled well and won themselves a place in paradise. But if there is no God, little is lost. 'Let us weigh the gain and the loss in wagering that God is. Let us consider the two possibilities. If you gain, you gain all; if you lose, you lose nothing.'

God in the future

Most observers are inclined to the view that the universe will expand forever. This is because of the dominance of the dark energy, which triggers an ever-accelerating phase. We are in a universe that is undergoing perpetual inflation. However, there is a minority view that holds that the dark energy may not be forever. The cosmological constant would not be constant, but could decrease in the future. In this case the acceleration stops and the universe could even be destined to collapse to a future Big Crunch.

This is not an especially heretical point of view, since we know that once, during the epoch of inflation, the cosmological constant must have been very large in order to drive the inferred acceleration. Now in a universe that consists only of matter, dark and baryonic, the fact that it is flat means that it is infinite and will expand forever. But this is a possibly misleading conclusion, since observations can never prove precise flatness. The universe need not be precisely flat. And its fate depends on the sign of the deviation from flatness. If the deviation from flatness erred on the spherical side – that is, the two-dimensional analogue of an infinite plane was replaced by the surface of a very large sphere – then the universe would eventually be destined to collapse. A Big Crunch would again confront us.

This has intriguing prospects. The universe could arise, Phoenix-like, into a reinvigorated and uncontaminated Big Bang. The US cosmologist Frank Tipler has exploited the philosophical and theological implications of a future Big Crunch. He argues that global anisotropies, exceedingly weak today as inferred from microwave background observations, are generically present at some infinitesimal level. In a future Big Crunch, anisotropies would be amplified and provide an infinite source of energy as the final singularity is approached.

This almost infinitely hot and hell-like future of Big Bang cosmology has been applied with logic, although some might also say with sophistry, to prove the existence of God, specifically a loving God who will resurrect us all to eternal life. Tipler claims that theology is simply a sub-branch of physics and that God is what he defines as the Omega Point.[1] This is the endpoint to our universe, when the future Big Crunch will provide an inexhaustible supply of energy (from the release of gravitational energy liberated by the collapse of the cosmos). Armed with that energy and a corresponding ability to store information, a supercomputer of the future, a.k.a. God, will attain unlimited power, resurrect the dead and bestow all sorts of blessings on humanity.

Most cosmologists are prepared to accuse Tipler of the direst crime, perpetrating a hoax of Piltdown Man proportions. Yet the reading public seems eager to lap up Tipler's theology. Even Wolfhart Pannenberg, an eminent German theologian, has spoken out in its defence. To provide a veneer of sophistication, Tipler has developed a mathematical framework that generates an almost impenetrable aura of erudition. Immortality, Tipler purports to prove, is an inevitable consequence of general relativity and quantum theory. It is when he starts using physics to prove the exist-ence of God that the scientist tunes out. But is dabbling with God any more insidious than dealing in time? And where does Tipler go wrong, if indeed he is guilty of the alleged crime?

Other researchers have already softened up the God-seeking audience. Physicists far more mainstream than Tipler have equated God with such fundamental entities as a set of equations or the Higgs boson, a hitherto undiscovered elementary particle. Particle physicists flock around 'Theor-ies of Everything' (which claim to explain the very basis of existence) like moths around a flame. Even observational cosmologists have entered the God stakes. George Smoot, the leader of the NASA team that discovered fluctuations in the cosmic microwave background, described his achieve-ment as seeing 'the face of God'. Paul Davies, seldom far from the fore-front of cosmology, has already written two books in which the casual reader might infer that he identifies God as a quantum cosmologist. Never one to be bashful, Stephen Hawking declared God unnecessary. Hawking proposes that the universe has no boundary in space or time, rendering a divine Creator superfluous. Even experimentalists have joined the party: the US particle physicist Leon Lederman talks of searching for the God particle.

Tipler takes a very different, personal tack that carries him into uncharted territory light years beyond the other God-dabbling scientists such as Paul Davies, Stephen Hawking, and their ilk. Tipler's theology, for instance, embraces not just cosmic structure but human sex: thanks to the Omega Point, it would be possible for each male to be matched not merely with the most beautiful woman in the world, not merely with the most beautiful woman who has ever lived, but with the most beautiful woman whose existence is logically possible. In this process, our bodies can acquire the most desirable characteristics, and unrequited love is certain to be requited. This astonishing vision, we are told, stems directly from Einstein's theory of general relativity.

Tipler's claims test the limits of how far science can take us in the ageless quest for an omniscient and omnipotent deity. Physics is far from confronting consciousness. For now, biologists laugh at the notion that quantum gravity could provide vital clues to the origin or evolution of life. Yet Tipler envisages a supercomputer of the future that will be able to resurrect human beings in full: our memories of passion, our thoughts of beauty, our dreams and desires. I concede that one must be exceedingly brave to make any predictions about the capabilities of supermachines billions of years in the future, but I cannot really believe what Tipler describes. My confidence in Tipler's grand predictions is not strengthened by the gaping holes in his more specific assertions.

The assertion that our fate in a collapsing universe will enable us to unlock the gates of heaven is flawed, for a simple reason. When the universe was one minute old, it was about as hot and dense as the center of the Sun. We are quite confident of this description, because of the remarkable success that the Big Bang theory has had in predicting the abundances of the light elements. These were synthesized in the intense heat of the first few minutes. The future evolution of the universe, were it to collapse, would return it to a similarly hot, dense state. No room there for any supercomputer, or for any recognizable being. Sex would not be much fun at 100 million kelvin. It seems that Tipler has constructed hell rather than heaven.

Life is likely to involve far more than a set of equations. I would contend that as many of the mysteries may reside in the uncertain boundary conditions and in the unpredictable behaviour at critical phase transitions as in the regime amenable to algorithmic computation. One may even have to go beyond physics to comprehend the complexities of nature.

Physics, for good reason, was born, and still resides in some circles, as natural philosophy. Indeed, physics and philosophy complement each other, in a relationship that deserves to be more than a mere relic from the amateur inquirers of that bygone era. Why should modern physicists give a hoot about philosophical issues? The reason, as I see it, is to keep the science in perspective. Interestingly, it is almost always the older, established physicists who turn their thoughts in the direction of the big questions of meaning and God. Steven Weinberg writes that 'The more the universe seems comprehensible, the more it also seems pointless.' Paul Davies urges that 'Science offers a surer path than religion in the search of God.' And Stephen Hawking offers his dream:

Then we shall all, philosophers, scientists and ordinary people, be able to take part in the discussion of the question of why it is that we and the universe exist. If we find the answer to that, it would be the ultimate triumph of human reason for then we would know the mind of God.

Such thoughts can be considered as complementing theology and need not be considered a substitute. Humility in the face of the persistent, great unknowns is the true philosophy that modern physics has to offer.

Note

1. See Frank Tipler (1994) *The Physics of Immortality*. New York: Doubleday.

20 Where Next?

Science moves, but slowly slowly, creeping on from point to point.

Alfred, Lord Tennyson

Prediction is dangerous. The Danish physicist Neils Bohr allegedly said 'Prediction is very difficult, especially about the future.' But this sort of view has never stopped us from speculating.

I expect the third millennium will bring in the long sought-after Theory of Everything, or something close to it. This is the Holy Grail of physics, a theory of matter and energy, of space and time, which will leave few questions unanswered or at least unanswerable. We will know how our universe began, and what there was before the Big Bang. We will understand how structure developed out of homogeneity over scales that range from infinitesimal to vast, from dimensions of 10^{-43} cm, where gravitation confronts quantum theory, to 10^{28} cm, where the world's largest telescopes confront the horizon of the universe. We will know how the fabric of three-dimensional space was formed out of higher dimensions, how the inexorable flow of time began, and how matter itself was created out of virtually nothing.

The Theory of Everything will be of little use in exploiting human potential or in predicting catastrophes such as volcanic eruptions, earthquakes, or the collision of an asteroid with the Earth. But by cracking the secrets of matter, it will yield the ingredients for unprecedented advances in technology, including domains as diverse as artificial intelligence, biotechnology, engineering, and chemistry.

One can imagine that the end of physics would then be in sight. If so, it is timely to reflect on what the physicists of the third millennium might do. One choice might be to become quantum engineers, tinkering with the Theory of Everything to extract and exploit every last nuance, and participate in the merchandizing of exotic applications that would help to ameliorate the problems of an increasingly polluted and crowded world.

Others, no doubt, would move on to the problem of consciousness, the ultimate barrier that distinguishes man from beast.

The reductionists of the future will not be happy with this, but neither will the theologians. Indeed, the theologians are still trying to assimilate the implications of the debate of Valladolid in 1550, in which the missionary Bartoleme de las Casas (1484–1566) tried to make the case for such a distinction.

Approaching omniscience

Computer power doubles every year. This has been a pattern that has been in place for half a century and seems destined to continue for a while. There is no reason to think that the rate of computational progress will slow in the next decade. One can just reduce the sizes of printed circuits and electronic chips down to multiatomic scales, a few billionths of a metre. Beyond that, one runs into quantum phenomena. In the quantum universe, prediction of our potential is more speculative. We may be hardware-limited unless some radically new direction is explored.

Quantum computing offers the promise of such a potentially unlimited future. Anything goes in the virtual universe, as long as we cannot actually observe it. In particular, quantum processes have multiple histories. Think of Schrödinger's cat, either alive or dead. That is the effect of only two parallel universes. However, the number in other situations can be multiplied a billionfold. And one can utilize this quantum multiplicity of universes to do one's computing, since all one cares about is the result.

A computer of sufficient power can simulate a man, or a woman. It could stitch a person together molecule by molecule. Indeed, a computer of the future could construct a superman, whose intelligence and capacities exceed our wildest dreams. So this leads to the ultimate question: why not God? What is the difference between the all-powerful supercomputer of the future and a deity? The answer can only be in something that is unique to human beings, something that beasts lack, something that a computer perhaps could emulate but could never create. This attribute is certainly not intelligence, nor is it beauty. Nor is it likely to be consciousness. One can imagine a computer that would be designed to be self-aware. A computer could be built to feel pain, and a computer could even write poetry.

But something is missing. Can a computer laugh? Can a computer undergo emotion? Can a computer fall in love? Or be angry? Can a computer achieve inspiration, that moment of eureka? No doubt a super-computer of the future, let us call it a hypercomputer, can appear to have such feelings. And no doubt a hypercomputer could be designed to be indistinguishable from a human being. Androids will be reality some day. There will be moments, however, when the reactions would differ between hypercomputer and human. This, one hopes, is the very essence of the difference between automaton and man. A future dominated by robots would not be worth fighting for. Nor would it be a very interesting environment. The peaks and troughs, the texture of human existence, would surely be absent.

Our omnipotent hypercomputer cannot be a man or a woman. The differences would be subtle but tangible. Yet there is no reason why the hypercomputer could not assemble a human being. This is biologically feasible. We are already capable of assembling artificial lungs and hearts after a decade of experimentation. Imagine the possibilities of biotechnology in a thousand years, or even in a million years, perhaps the briefest instant of time discernible on the cosmic clock that marks the ageing of our Sun.

Every trace of genetic material could be manipulated to reproduce anyone who has ever lived. One could clone a man, as long as one had some master template in memory. This technology is already useful for prolonging human life, and we have barely begun. All body parts would be replaceable. To the extent that all of our memories are electrically recorded on to neurons in the cerebral cortex and therefore reparable and transferrable, replacement of the brain should be equally feasible. Making a man should be child's play to our hypercomputer.

So why isn't the hypercomputer a substitute for God? Poetry provides an answer. A computer can copy and even modify genetic patterns, but it lacks the discrimination to generate a human genius. Its products may have unsurpassed intelligence, but this is not sufficient to create the genius of a Shakespeare, a Mozart, or an Einstein.

Imagine taking a tribe of monkeys. Put them to work at a row of typewriters. With enough monkeys slaving away, sooner or later one will produce a play such as *Othello* or *Macbeth*. It is incredibly improbable. If only the monkeys could learn by trial and error. None the less, it is possible, in principle. The odds are greatly reduced if the letters typed in by the monkeys are not completely at random. Patterns can emerge, due, for example, to the design of the keyboard. This will speed up the process, but it will still take almost an eternity. This does not validate the random selection process, by which one would characterize computer exploration of a difficult task, trying all possible permutations until one succeeds. This is how IBM's Deep Blue computer defeated the world chess champion, Gary Kasparov in 1997.

One has to add a spicing of human intervention. One needs some means of distinguishing the true *Othello* from the innumerable failed *Othellos*. This is beyond the capability of any monkey, or any computer. It requires human judgement, or even inspiration. Here is another example. Take the same tribe of monkeys. This time equip them with oil paints and canvases. Let them loose. Remarkable abstract oil paintings occasionally emerge, the equals of any Jackson Pollock. But these are transient phenomena. If the paintbrush is not judiciously removed from the monkey's paws at precisely the right moment, a black morass results. And who judges when to remove the brush? Again, no monkey, and indeed no computer, could do this with the aplomb and creative flair that, say, Jackson Pollock himself might have done, were he running the experiment.

There is seemingly a mystery here, just as there is in the apparent macroscopic reality of certain quantum phenomena. Roger Penrose has argued that part of the mysteriousness of consciousness may have a quantum physical origin. The US philosopher Rick Grush comments that 'It is not clear how a collection of molecules whose chemical composition is not unlike that of a cheese omelette could be aware of anything, to feel pain, or see red, or dream about the future.' But artificial intelligence studies suggest that it may not require miracles to comprehend consciousness, just enough knowledge, coupled to memory and decision-making skills, which a hypercomputer could no doubt acquire.

Quantum gravity is not as remote as might be thought. Look up into the sky with microwave vision. The temperature fluctuations on large angular scales cannot have been generated by any causal process that operated since the universe underwent inflation at 10^{-35} seconds after the

Big Bang. Our best explanation for these fluctuations is that they are quantum fluctuations that have been amplified on to macroscopic scales by the process of inflation. At this moment, corresponding to when the fundamental forces were last unified, the universe must have inflated exponentially by a factor of at least 10^{60} to be able to expand subsequently to its present size. As a consequence, the once causal quantum fluctuations, no more than 10^{-35} light seconds across, were imprinted on to scales of billions of light years. As we study the cosmic microwave background, we are viewing quantum fluctuations in the sky. This is a discovery that has revolutionized modern cosmology, for it provides the explanation for the seeds of large-scale structure formation.

Quantum gravity has far more speculative implications that border on science fiction. The taming of black holes and wormholes constitutes one of the most exciting prospects that a nearly infinite universe could hold for us. Our energy crisis would be resolved. Unlimited knowledge could be tapped. But this comes at a price.

The classical view of a black hole is marred by one ugly feature. At the core of the black hole there is a singularity. This is a forbidding concept, since literally all hell may break loose should one get too near the singularity. Some cosmologists are convinced that such a singularity is never accessible, or naked: it is always shrouded by the black hole horizon. In this way we can live our lives without undue fear of a confrontation with the horrors of facing a singularity, with the inevitable breakdown of the physical laws that govern our existence and even our sanity.

Were a naked singularity to be found it would immediately allow the possibility of extracting unlimited resources from other universes. Miracles could be performed. Time travel would become feasible, since space and time reverse their roles. One could travel in time, either far into the future to escape any of the unfortunate calamities of our current era, or into the past, to pursue our dreams of long-lost Elysian Fields. It is this prospect that horrifies some, for one could, if sufficiently perverse, go back in time, seek out one's grandmother in her infancy, and murder her in order to challenge future generations of physicists. For now our notions of causality would be overturned: the impossible is possible, there is a fundamental contradiction in the laws of physics.

Quantum gravity rescues us from this quandary. Such adventures into the past are subject to the laws of quantum uncertainty. The probability of actually finding a particular individual at a particular place and time would be vanishingly small. One's ancestors are safe.

Some cosmologists are inclined to think the game is almost over. Stephen Hawking has written that 'There are grounds for cautious optimism that we may now be near the end of the search for the ultimate laws of nature.' And more strongly: 'The most important fundamental laws and facts of physical science have all been discovered, and these are now so firmly established that the possibility of their ever being supplemented in consequence of new discoveries is exceedingly remote.' But even Hawking admitted that something was missing.

Even if there is only one possible unified theory, it is just a set of rules and equations. What is it that breathes fire into the equations and makes a universe for them to describe? The usual approach of science of constructing a mathematical model cannot answer the questions of why there should be a universe for the model to describe. Why does the universe go to all the bother of existing?

I am more inclined to think we are just at the beginning of our search. Henry David Thoreau wrote that 'If you have built castles in the air, your work need not be lost; that is where they should be. Now put foundations under them.' New ideas are needed, that is sure. But they are destined to meet resistance. Einstein was only too aware of this: 'He who joyfully marches to music in rank and file has already earned my contempt. He has been given a large brain by mistake, since for him a spinal cord would fully suffice.'

I am optimistic for the future. Great progress has been made in the past. There is a niche for humanity in the vastness of the universe. And the nearly infinite universe that our cosmology favours has immense wealth in store for us. One cannot hide a sense of wonder and awe at the vastness and beauty of the cosmos. Great surprises lie ahead, some of which we will surely glimpse with the aid of the ever larger telescopes and more powerful particle accelerators that are being planned over the next decade or two. The future beckons brightly.

■ INDEX

Italic numbers denote reference to illustrations.